# SCREEN SCHOOLED

Two Veteran Teachers Expose How
Technology Overuse Is Making Our Kids Dumber

## Joe Clement & Matt Miles

Published by Black Inc.,
an imprint of Schwartz Publishing Pty Ltd
Level 1, 221 Drummond Street
Carlton VIC 3053, Australia
enquiries@blackincbooks.com
www.blackincbooks.com

Original edition published by Chicago Review Press Incorporated,
814 N. Franklin Street
Chicago IL 60610

9781863959995 (paperback)
9781743820193 (ebook)

 A catalogue record for this
book is available from the
National Library of Australia

Cover design by Mark Whitaker / MTWdesign.net
Typesetting: Nord Compo

Printed in Australia by McPherson's Printing Group.

# Contents

# Authors' Note

To PROTECT THE IDENTITY OF STUDENTS, parents, and colleagues, some names and minor details have been changed. For ease of reading, we've written the book in one voice and one point of view, but it was written together in full partnership. It represents a composite of both authors' ideas and experiences. Finally, we should note that the opinions expressed in this book are our own. We are not in any way speaking on behalf of our employer.

We would like to encourage parents, educators, scientists, and others to engage with us about the critical issues discussed in this book. (We recognize the irony of using technology to connect with readers, but would categorize this as an example of the productive and collaborative uses of technology we're seeing too little of in schools today.) Please visit us online at www.screenschooled.com to learn more about us and the book, at www.paleoeducation .com to learn more about the research driving this movement and about ways to teach and learn simpler, and at our Google community "Beyond the Screens" to share ideas and resources with like-minded parents and professionals.

# Introduction

MAKE A LIST OF TEN THINGS that kids today need. Go ahead. I'll wait.

Thanks for doing that. Now look at that list. Does it include "more screen time"? I didn't think so. Mine doesn't either. If I had asked you to make a list of twenty things kids today need, would "more screen time" have been on that one? How about if it were a list of a hundred or a thousand things that kids need? The point is that while there is nearly universal agreement that kids today do not need more time on screens, schools are doing what they can to make sure kids spend more time on screens.

Matt and I have been teaching for over thirty combined years. We have taught thirteen different subjects to students in six different grades. We coach three different sports and routinely interact with students all over the age, socioeconomic, racial, and social spectrums. We also have kids of our own, ranging in age from one to eighteen. When it comes to kids, we certainly have not seen it all, but we've seen a lot. We have also interviewed dozens of colleagues, scientists, and politicians and have read everything we can about the intersection of digital technologies and the way kids learn. We are in a unique position to share with parents what is happening in schools when it comes to screen usage and how it's affecting our kids. In short, it's not good.

You may well be thinking, *These are just two curmudgeonly, angry teachers railing against these "young whippersnappers and their newfangled contraptions."* Not exactly. While we are teachers, we are neither curmudgeonly, angry, nor antitechnology. I'd like to think we're too young to be curmudgeonly, and we love teaching far too much to be angry. As far as our comfort with technology, I was a UNIX system administrator before becoming a teacher. Matt was an IT major in college before a last-minute switch to education. He is his department's "technology representative" and teaches other teachers how to use the school's grading, test creation, and attendance software. We are more proficient in Excel and Access than any teachers we know. But several years ago, the two of us began to question a notion that we were hearing with greater frequency—that more screens and technology should be incorporated into young people's education. This idea was repeated over and over in educational circles, so much that it just became accepted as truth. Schools all over the country began spending billions of dollars on educational technology and requiring teachers to incorporate it into their classrooms.

Before this time, educational technology, or "ed-tech," was something used in the background of education, designed to help solve administrative problems. Old paper gradebooks were replaced with digital ones. These programs did the calculations for teachers and made posting grades much easier. Attendance was also digitized. Teachers could mark a student absent, and an automated phone call home could be made with the click of a button. Then schools started adopting teacher webpages where all the digital materials used in a class could be posted. Now students could access materials they either missed or lost without having to ask the teacher for copies of their notes and handouts. All of these changes made the lives of the teachers, administrators, and students much easier and didn't interfere with educational goals.

But by the mid-2000s ed-tech's focus became more student centered. Increasingly, lessons and instruction were digitized and automated. Schools began spending heavily on laptops and iPads. Online textbooks replaced paper ones. Teachers were trained in the latest and greatest educational software. We were shown how to incorporate YouTube clips into presentations, set up web-based scavenger hunts, and replace our old, boring PowerPoints with "Prezis" (essentially PowerPoint with more spinning and movement). When students found these projects cheesy (because they were still educational), teachers were encouraged to kick it up a notch. "We need to meet kids where they are" is a phrase echoed at every ed-tech presentation. Now teachers are encouraged to use laptops and iPads in every class. Instead of introducing education through educational software, teachers are now struggling to cram education into the technology with which young people are comfortable, like social media and video games.

But with all this talk about "paradigm shifts" and catchy slogans like "Meet kids where they are," everyone was glossing over the most important question: *Is this what is best for students?* Educators have become so consumed with *how* to incorporate more technology into the education of young people that they forgot to ask, *Should we do this?* Ed-tech firms, with their large marketing budgets, had convinced parents and educators alike that their products are necessary for future student success.

However, these messages contradicted everything Matt and I were seeing in children. It seemed obvious that students weren't benefiting from the technology in their lives. Rather, they were quickly transforming into technology addicts. They were consumed by video games and social media. Technology seemed to be shaping their social interactions and even how they thought. I decided to take an in-depth look at the claims of ed-tech firms. I noticed that too often there wasn't any actual evidence supporting them. There were too many questions that were met with glittering generalities and nonanswers. So I decided to see what unbiased social scientists and neurologists had to say on this topic. I make use of this literature review in the following chapters.

The problem that I've discovered, and will continue to outline throughout the rest of the book, is that kids spend too much time passively consuming entertainment forms of media on their screen-based technologies. The results of this problem are that students struggle to focus, critically think, problem-solve, and interact socially. This problem has also had a negative impact on families, as well as the mental health of kids. This problem is now ingrained in our society, as seemingly everyone has accepted this overindulgence of technology as the "new normal." It starts in the home, where many parents are allowing screen use for upward of eleven hours a day. But it's also now being fueled by schools, which have encouraged this way of living by going digital.

That is my experience. I hope that by the end of this book you will agree with me. I want to be clear at the outset, though, about what I'm *not* saying. I'm not pushing to remove all screens and digital technologies from schools. That would make no sense. How can kids learn computer programming without computers? How can they learn to efficiently work with data if they cannot use spreadsheet and database software? Some teachers excel at incorporating technology into their lessons in creative and engaging ways. It suits them and their style and augments specific lessons. I say more power to them. It certainly isn't my contention that all technology is inherently bad. Some technologies have made the impossible possible in education. One colleague uses video conferencing software to have virtual "Socratic webinars" with students and leaders from all over the world. There's no denying that's amazing. However, this designation of "revolutionary potential" is too often slapped on everything that has a screen and is sold to parents and educators. Most of it is not amazing, and much of it is less effective than traditional forms of instruction.

Finally, my intention isn't to blame parents and educational policy makers. I'm confident in saying the overwhelming majority of parents and educators are trying their best to do the right thing for kids. My goal is to challenge people to open their eyes to what is happening to our children. Technology is no longer just an important part of young people's lives. Technology has become their lives. It consumes them in every conceivable way. I question the notion that the older generations of parents, educators, and employers should simply accept this as a new reality. I question the notion that schools should shift their focus from teaching in the best possible way to being sure to teach with technology. I question the claims of technology advocates that knowledge is no longer necessary in the digital age, in which being able to look things up is as good as actually knowing things.

What I aim to do in this book is two things: teach and empower. I want to teach about the changing relationship between schools and screens, how schools and kids are changing because of the overuse of screen time, and why this is bad for our kids. I then want to empower you to do something about it. I provide you with practical steps you can take today, next week, and next year that will help your kids and your kids' schools navigate this thorny issue. Over the years, I have sat in enough conferences and taught enough classes to know what real help from parents looks like. It will not be easy. But our society is at a crossroads. I want you to join me on the road to mental, emotional, physical, and psychological health for our kids. What it takes is a critical mass of parents, teachers, and students demanding change. We can do this.

# 1

# These Kids Today

*Kids today are being controlled by smartphones,*
*and becoming enslaved by them.*
—Ryuta Kawashima, professor at the Institute
of Development, Aging, and Cancer at Tohoku University

SOMETHING IS NOT RIGHT with today's kids. You know it, and I know it.

You know it because you are seeing it in the morning and in the evening. You have a gnawing suspicion that they're not quite as physically active as they should be. Maybe they have trouble with routine social interactions. They sometimes struggle to solve life's most basic problems.

You might be thinking, *I provide and do everything I can for my children at home. I wonder if something is going on at school.* You don't really know what happens during the bulk of their day. That's where I come in. I do know what happens, and you're right—it's not good. I'm not going to sugarcoat the truth. What makes this warning different from others about "these kids today" is that I believe the problems many of our kids are having with their academic, social, and family lives are being caused in part by their schools, which are doing the bidding of the government and corporate America. There are, of course, many factors that cause kids to struggle with things like focus, critical thinking, problem-solving, and social interaction. I'm convinced that the largest single factor, however, is overuse of screen time, both in and out of school.

Like most parents, you're likely ambivalent about screen time for kids. You feel a little hypocritical setting limits on it for your own kids when so much of your day is spent plugged in to your phone, tablet, and laptop. You may hear commentators like social media scholar danah boyd (who does not use capital letters in her name) argue that kids today need fewer parental restrictions on their digital lives and more freedom to explore the online world and social media (she works for Microsoft Research). To you that sounds extreme, but you want your kids to be well-liked and to have the same things other kids have. You don't want to be "that parent," the only parent in the neighborhood who won't get your fourth grader a cell phone; and you don't want your child to be "that kid," the only kid in the neighborhood without one. And you like the idea of being able to keep in touch with your children, especially in case of emergencies.

However, you know the temptation to misuse a phone is a lot to put on a young person. You would prefer it if your kids had other kids actually, physically, come over to your house and do things without screens, like you and your friends used to. You know the Internet is a sewer of amusing cat videos, video games, and pornography, but it also enables communication in ways unthinkable only a few years ago. You realize that schools can use this amazing potential to help kids learn incredible new things in ways that weren't possible when you were their age. But you also realize the tremendous potential for wasted time. You know that "social media" has the potential at once to connect and isolate us.

You recognize the fact that giving your child a tablet sure makes him quiet, and quiet is nice. However, part of you also misses the time in the car, before built-in DVD players, when you would play the alphabet game to pass the time. Entertaining and occupying your kids—especially during summertime—is hard work. Screen time makes it very easy, and if you allow your kids access to only educational apps and games, what could possibly be the harm? The fact that you're reading this book means that you recognize there is harm. You're right to be concerned.

When I speak to teacher and parent groups, I inevitably get a question that goes something like this: "How is this any different from when television was invented and the old people said it would rot our brains? It didn't. We're fine." That's a great question. However, saying that the advent of television and smartphones is similar since they both have screens is like saying that lightning bugs and lightning are similar because they both give off light. Especially early on, television and television shows were (and in many cases still are) events. Families and friends gathered to experience shows together. More important, televisions were in the living room, and they stayed there. These two characteristics are what make the advent of smartphones a new chapter in human history. The smartphone, tablet, or laptop is made to be used and experienced alone. Each is a table for one. Further, they go everywhere with us. We no longer have downtime, we no longer have to wonder about things, and we no longer have to be bored. Those three things sound great to someone who is not thinking very deeply about the human experience. However, downtime is necessary for the human brain; wonder is what results in human advancement, and boredom results in creativity.

Many excellent books today explain why we need downtime, wonder, and boredom and how screen time is impinging on all of those. *Mind Change*, by Oxford neuroscientist Susan Greenfield, explains that throughout human history the brain has changed as its environment has changed. She points out that the amount of screen time we consume today, and the way in which we consume it, constitutes profound changes in the brain's environment. Therefore the brain must be changing. These are the main questions at this point: *How* is the brain changing, and are those changes desirable? From two educators' perspectives, young peoples' brains *are* changing, and these changes are affecting their ability to learn. For years, I have seen this decline firsthand in the classroom and no longer have any doubt about it.

## We Don't Always Have to Meet Kids Where They Are

The fact is that schools pushing ever more screen time on kids often say this is because educators need to adapt to the world in which students are living. However, schools don't need to cater to every behavior students indulge in. For instance, teenage smoking used to be a huge problem in America, much bigger than it is today. At many high schools in the 1970s and '80s, plumes of cigarette smoke billowed out of student restrooms, a symptom of this problem. Many students could not get through the school day without a cigarette. How did schools respond? They provided a place in the school for students to go smoke during the school day—a "student smoking court." Imagine that; everyone knew cigarettes were very bad for you, especially for kids. In spite of that, schools actually provided a place for children to smoke! Taxpayer dollars went to support teenage smoking. That actually happened. Thankfully, schools eventually realized how dreadful that was, shut down the smoking courts, and educated kids on the dangers of smoking. The result? The CDC issued a press release in 2014 saying that teen smoking was at its lowest point in decades. Lesson learned.

Rewind. Why did schools provide places for children to smoke? Because administrators thought, *Students are going to smoke, so we should learn how to live with it and make the best of the situation.* Sound familiar? Students are doing something that we know is not good for them, and society—schools included—is complicit in perpetrating the damage. Think of just a few of the ridiculous things you did in middle school and high school, those things that your parents and teachers told you were bad and you needed to stop. Perhaps your friends did some of these things as well. Now that you're an adult looking back, should your parents and teachers have said, "Well, if all the kids are doing it, we should just get on board?" If you're like me, the answer is a resounding *no*.

## Meet Brett, the digiLearner

How much do you know about what goes on from the time your son or daughter leaves your home in the morning until he or she gets home in the afternoon or evening? If you're like most parents, not much. Teachers are the ones who know your child's school behavior best. And we've watched firsthand as young people have been profoundly changed by their technology, seemingly overnight. What follows is a fairly typical day in the life of a modern high school student. I will call the student Brett. He is a fictional character, but he is absolutely based in reality. He is an amalgamation of observations of students over the last several years and of stories I have heard from parents, other teachers, and students themselves, and on my observations of my own children.

Brett is sixteen years old and today is a school day. It is 6:05 AM and the alarm on his cell phone rings. He reaches under his pillow, where he keeps his cell phone at night, and automatically navigates to the alarm function to put the phone to sleep. After nine more minutes of blissful rest, the alarm sounds again. It is now 6:14 and he knows he must get out of bed. "Brett! Please get going," his mom calls from downstairs. "We have to be out the door in thirty-five minutes, and you still have to shower and eat!" With a moan, he lifts himself out of bed, grabs his phone, and checks for any messages he may have missed in the middle of the night. Brett's time of being "unplugged" from the web and social media has ended for the day. He feels a warm glow as he sees that he got two "likes" on his 1:15 AM status update, "Going to bed."

He staggers into the bathroom and turns the shower on. While he waits for the water to warm up, he takes stock of himself in the mirror. His pale skin is nearly blinding in the harsh light of the bathroom. His arms are twigs protruding from narrow shoulders. His midsection is doughy and his legs bony. Long gone are the scrapes and bruises that told the story of a child who loved to run and play outdoors.

After his shower Brett returns to his room. Before getting dressed, he checks his phone. A friend has sent him a link to a hilarious YouTube cartoon video featuring a homicidal llama, as well as a Snapchat picture of himself with a Mohawk combed into his wet hair. After watching the video, Brett pops in his earbuds and heads downstairs. He is exhausted from his late night of homework and needs his morning music to get going. As he passes by, his mother gives him a few instructions. Earbuds in, Brett hears only the words "Pop-Tart" and "bathroom." He goes into the kitchen, grabs a strawberry frosted Pop-Tart and a Capri Sun. On his way out the door, he nearly stumbles over his mom. She motions for him to remove his earbuds. "Didn't you hear me?" she asks. He replies, "Yes I heard you. I told you I can multitask. I can listen to music and still get things done. I got my Pop-Tart, and I don't have to go to the bathroom, so let's just go." She informs him that she asked him to *not* eat yet another Pop-Tart for breakfast and also to turn off the light in the bathroom, which he had left on. Brett shrugs, drags himself upstairs to turn off the light, and heads out the door to the car.

To an outside observer, the ride to school is silent. However, this observation misses the many things Brett is doing with his phone. He checks Twitter and finds that two of his friends are also heading to school and dreading it (#SkoolSux). He and a friend exchange two more hilarious YouTube videos involving cats. He snaps a picture of himself with his head out the window of the car, and sends it to a friend. As they enter the line for kiss 'n' ride, his mother tries fruitlessly to engage her son in conversation. She looks at him, glued to his screen, and figures that it is easier to just let him have his last few minutes of peace before a hard day at school. *Kids today have it so much harder than we did,* she thinks to herself. As he gets out of the car she says, "I love you, honey." Eyes still glued to the screen, Brett replies, "Love you, too." He closes the door behind him and walks into school.

Brett's mother takes a minute and watches him disappear into the sea of kids funneling through the school door. She can't help wondering what is happening inside the school. Buses, student drivers, and other harried parents are careening every which way as she navigates her way out. As she hits the highway, she turns her mental attention to a presentation she has to give later that day, and thoughts of Brett's day fade into the background.

At work, though, a nagging feeling returns. She sends Brett a quick text, asking how his day is going. Texting is her chance to enter her son's world. It's obviously not as good as a real conversation and a hug, but it's better than nothing. As she hits send, she's still wishing she knew more about what is happening at school.

I can fill her in. Let's go back to the morning drop-off.

Brett gets out of the car, without taking his eyes off the screen. "Love you, too," he says, closing the car door behind him. As he enters the building, all around him kids are scurrying to class. He does not truly see any of them. Some are also glued to their screens, while others are talking to one another. Some students are furiously finishing homework, and others are clumsily engaged in rudimentary courtship rituals as the seconds tick down before class. Brett finds his friends standing near their lockers. They are all staring into their screens. They're talking, but not to one another. They are proudly narrating what is happening in the games they are playing. Brett joins in until the warning bell rings. The party breaks up and Brett slips into his first period class, English. His teacher asks him to take out his earbuds. He silently complies.

Brett now must go into "stealth mode." This is the part of the class where many students slump in their seats with their cell phones in their laps, hidden behind their desks. This would be an excellent technique if the vacant downward stare and slight glow from the screen didn't give them away. The teacher now has a choice. She can ask Brett and others like him to put the phones away or she can rationalize her decision to ignore it. *It's their education. If they'd rather play* Words with Friends, *then fine—at least they are building vocabulary, or, I know they'll put the phones away in a few minutes. It's still early and we'll just ease into class today.* Little does she know, they stopped playing *Words with Friends* two years ago. They now learn important new skills by having a monkey shoot balloons. Today she chooses to rationalize, thinking, *I'm tired of fighting this, and most of them are doing just fine anyway.*

The teacher asks students to pass in their *Hamlet* papers, which are due today. The girl next to Brett nudges him and repeats what the teacher just said. Brett does not have his. He had completely forgotten it was due today. He checks the calendar on his phone. Sure enough, he had entered it, but forgot to attach an alarm reminder. Panicked, he questions the teacher about when this due date was announced. The teacher informs him that it was announced three weeks ago, at the beginning of the unit, and she reminded the class last period. That doesn't really help Brett because he spent most of last class on his phone, watching *The Campaign* starring Will Ferrell. It was hilarious. Two of his more studious classmates remind him that it has also been posted on Blackboard (their class website) for the last three weeks. Brett protests further, asking why the teacher had not e-mailed a reminder to everyone. She curtly informs him she sent everyone a message on the Remind app she uses for the class, which would have sent a text message directly to his phone. However, he never downloaded the program, which promises to make classroom communication easier. He thinks it's dumb. Why would he want text messages from his teacher about homework? That's creepy.

He then notices two new text messages on his phone. One is from his mom. "Nothin," he replies to her question about how his day is going. The other is from another student in class, two desks away: "4get it dude. She a b1tch." Brett smiles to himself and cuts the protest short. *Great,* he thinks, *just one more thing to add to the pile of work for tonight.* He quickly glares into his phone, takes a selfie, adds the caption "Soooo pissed!" and sends it via Snapchat to a few friends, including the one two desks away.

Because Brett is not failing any of his classes, he next has a study hall. He can use this period in virtually any legal way he sees fit and can go anywhere in the school to do it. As he does every day, he goes into a classroom where several of his friends are. The room is packed, but eerily silent. A few students have books open and are reviewing for tests later that day. However, most students—Brett included—are on their devices. Of these

students, most have their earbuds in. Brett is watching another
movie he has downloaded to his phone. The movie is interrupted
periodically by texts and Snaps from friends. He texts his mom
back with the emoji of a smiley face licking its lips (she asked
him if he was OK with pork chops for dinner). After forty-five
silent minutes, the students leave wordlessly. Brett is now off to
history class.

As the period is beginning, Brett's attention turns briefly to
two students trying to impress the teacher by reciting all US pres-
idents in order. They stumble on the thirteenth president. As they
debate whether it is Buchanan or Pierce, Brett takes a break from
the video game he is playing on his phone. He goes to Google
and discovers the identity of the thirteenth president. "Millard
Fillmore," Brett blurts out. "Nice work, Brett," the teacher chimes
in. "He took over the presidency after Zachary Taylor died of . . ."
Brett loses interest and returns to his game, feeling pretty happy
that he knew something the others did not. Using his phone,
he finds a picture of Millard Fillmore, does a face swap using
Snapchat, and sends it to several friends with the caption, "Brett
Fillmore, bitches!"

Later in the day, in Spanish class, students are reading aloud
from an article in a Mexican newspaper. Brett has found the article
online and is filtering it through Google Translate to help with
unfamiliar words. However, he also finds some amusing bad trans-
lations. He stifles a laugh when he sees that the Spanish for "early
childhood vaccinations" has been translated as "unusual tomato
farmers." It is at this moment that his teacher calls on him to read.
Brett asks where they are in the article. His teacher is annoyed
that Brett was not paying attention. Brett apologizes, but explains,
"I found the article online and was translating it. That's why I
wasn't paying attention." He feels hurt and disconnected from this

teacher, who simply doesn't understand about how he learns and how well he can multitask.

In the day's final period, physics, Brett works on a lab report. They have been studying Newton's second law of motion (which boils down to one of the fundamental concepts of motion in the physical world: force equals mass times acceleration). The teacher is allowing students to use the class computers to complete some background questions. Brett has had some trouble with this unit and hopes this computer time will help clarify things for him. When he gets the laptop, Brett immediately and instinctively goes to Google. He wants some background on Newton's second law of motion. He types "Newton's law" into the search box. The first three links are about Newton's first law of motion. Brett quickly surmises that this means there is nothing on the Internet about Newton's second law. He approaches his teacher and delivers the bad news, "I looked on the Internet and there is nothing about Newton's second law. Only the first law." Incredulous, Brett's teacher replies, "You're telling me that there is absolutely nothing on the Internet about one of the most important, fundamental laws in all of physics?" Brett says, "I know. I couldn't believe it, either. I guess I can't complete the lab report," and he returns to his seat, satisfied that he has done all he can with this assignment. After a quick search of his own Brett's teacher goes to him and says, "Brett, I came up with almost three quarters of a million hits when I searched just now." Brett is back on his computer, watching a thirty-year-old man wearing a headset play a video game. Videos like this are blocked by the school's Internet firewall. Like most of his classmates, though, Brett knows how to get around it. Without looking up, he tells his teacher he will keep looking. At that point, two other students have questions for the teacher and he goes to assist them. Brett uses his phone to take a selfie of himself "sleeping." He sends his eighty-seventh Snap of the day.

He has received over a hundred. He also replies to his mother's third text: "I love u 2."

After the final bell, Brett meets three friends for the afternoon carpool home. They pile into a waiting minivan driven by one of the other parents. On the ride home, the boys make a plan to meet up and play their favorite MMORPG (massively multiplayer online role-playing game). "Meeting up," of course, does not involve actually being together in the same physical location. It means each boy going to his own home, logging into the game, and finding one another online. Once home, Brett grabs a soda and settles into his gaming chair. It is a beautiful day outside, a fact lost on Brett. There are zombies attacking the city, and he must help defend it. He quickly locates his three friends, and the battle begins.

Brett's mother soon arrives home from work and calls out to Brett. She gets no response so she checks the family room. Brett is there, fully engaged in his game. He is wearing a headset and directing his forces. She watches for a minute, trying to figure out how to draw him into conversation. She eventually gives up.

Later, the family convenes for dinner. Brett's earbuds are not in, but his phone is right next to him. As his parents talk, Brett is responding to texts from one of his study hall friends. Slightly annoyed, Brett's mom asks him to put the phone away. "Why do you always want to stop me from connecting with people?" he asks. To Brett, texting friends and playing games online is social interaction. His father asks why Brett no longer invites friends over to shoot baskets or go exploring in the woods. Brett explains that those things do not interest his friends. Brett says—at least in part to please his parents—that he would like to do those things, but that's just not what kids do these days. "We just like to hang out," he says. Brett's folks have seen Brett and his friends "hang out" before. "Hang out" seems to be code for "sit in a room together, not really talking, and staring at cell phones." Brett has a point. What is he to do if his friends only like spending their time in front of screens? Brett's mother thinks about how different Brett's relationships seem from hers.

Brett stares down for the rest of dinner, feeling at once isolated from his friends and his family. After dinner it is time for homework. Brett's textbooks are all online, so his homework must be done on the computer. He heads to his room and closes the door. Brett logs into his online math book. He also logs into his accounts on Snapchat, Twitter, Spotify, and Instagram. Ten math problems take him two hours: twenty minutes to do the work, and one hundred minutes to interact with his social media and music. He does not like that it takes this long. However, his friends all expect him to respond immediately when they post something amazing . . . and *everything* they post is amazing. Further, Brett feels these breaks help with his anxiety. He always has so much work that he can never seem to get caught up. Some days he feels so anxious and overwhelmed that he cannot get out of bed. Of course, missing school means falling further behind, so he typically soldiers on. The only thing that makes the homework bearable is the interaction with his friends. At 9:30 PM his mother goes in to check on him. Brett's back is to her. His earbuds are in, but the screen shows his online physics book. *Poor, hardworking kid,* she thinks to herself as she closes the door without disturbing him. *Four or five hours of homework every night. I don't know how he does it. I never had that much work at his age.*

It is now time for Brett to turn his attention to that *Hamlet* paper. Eventually, he takes out his earbuds, brushes his teeth, and crawls into bed. He checks Snapchat and then updates his status one last time: "Time for bed. Today sukt." He closes his eyes. In about six hours, he'll do it all again. His digital clock says 12:15 AM. It is a digital world, and Brett is a *digiLearner*.

I know Brett well. I also know Brianna, Brett's female counter-part. I teach more and more students just like them every year. As many educational-technology advocates claim, they have been profoundly shaped by the technology that has come to dominate their lives. As an educator, however, I can no longer pretend these changes are good.

2

# The Myth of the Technology-Enhanced Superkid

*Young people's engagements with digital technologies*
*are varied and often unspectacular—in stark contrast to*
*popular portrayals of the digital native.*
—Neil Selwyn, professor at the Institute of Education
at the University of London

WHERE ARE THEY? Where are the technology-enhanced superchildren we were promised? We as a society have been inundated with claims that our children's use of technology will lead to improved multitasking, increased critical thinking, greater social awareness, sharper memories, and, in general, a generation of kids more capable than any generation that came before. These claims have dominated discussions within education circles for years. Now, they've spilled out into the mainstream and have come to dominate how we view the role of technology in parenting, coaching, or anything that involves interaction with kids. Dare to question these claims and you can easily find yourself on the outside looking in. You'll be painted as a Luddite—a stick-in-the-mud who fears change. But as I will discuss in this chapter, these claims are not only unrealized, they're entirely false.

## Educating the Teacher

Whenever a group of high school kids are huddling in a corner of the room, crouching together around a single cell phone and giggling, it's almost never a good thing.

It was the school's free period, where students could use the time however they saw fit. Although it was intended for studying, making up missing work, or talking to teachers, most kids used it to listlessly stare at their tablets and phones. A group of four students, two boys and two girls, who had come into my classroom, was doing just this. So these kids' high-pitched giggles echoed throughout my otherwise quiet room. I could tell by the restrained tone of their voices and guilty body language that they were looking at something they shouldn't be.

Fearing the worst, I decided to check on them to make sure everything was on the up-and-up. But they were so entranced by what they were watching, none of them noticed me as I approached their group. As I got within a few feet of them, I could hear shrieking coming from the speaker on the phone. *Oh no,* I thought to myself. As I got within striking distance, I asked, "*What* is that?" They jumped. The boy holding the phone, Isaac, tried to cram it back into his pocket.

"Nothing," he answered.

"Something for math," one of the girls, Monique, contradicted.

"You were giggling at a math video?" I asked, just to let them know I wasn't born the day before.

Without looking up from her screen, an older girl sitting next to them ratted them out. "They were watching a PewDiePie video."

She somehow managed to roll her eyes, indicating her annoyance with the group, without taking them off the movie she was watching.

"Pew Die Pie, who's he?" I asked, exposing my ignorance. When I remember this moment, I hear a record scratch in the background as all the other students in the room put down their phones and turn in disbelief. They looked bewildered. Who doesn't know who PewDiePie is?

"'PewDiePie,' it's one word," replied the other boy, Liam. "And it's pronounced '*dee*,' not '*die*.'"

Knowing the cat was out of the bag, Isaac explained, "He's a YouTube star. He's become a billionaire by posting videos of himself playing video games."

After the initial relief of *Thank god, they weren't looking at a torture video* wore off, I realized how odd that statement was. "Wait," as I paused for a moment to choose which one of the million questions that was coursing through my head I would ask first, "how do you become wealthy off YouTube videos?"

"He's only worth $100 million, not a billion," responded Liam with information he had just Googled on his phone.

"Companies pay him to review video games, plus he makes money from ad sales based on the number of subscribers he has on his YouTube channel," explained Monique.

"Reviewing video games, is that what he's doing?" I asked. "It sounded like a collection of fart noises and pig squeals."

"Yeah, it's hilarious," Isaac said, "He makes funny voices and adds commentary when he's playing games."

"The game he's playing, do you have it on your phone?" I asked.

"Yeah, it's Minecraft," answered the second girl.

"Why would you watch someone play a game when you could be playing it yourself?" I asked. This seemed absurd to me. I thought back to video game playing days of my childhood.

There would be ten of us gathered around a single television in the house of the one kid in the neighborhood with the latest and greatest game. One or two of us would be in a momentary state of euphoria as we enjoyed our brief chance to play. The other eight or nine sat there writhing in envious misery, openly rooting for the player's immediate demise (in the game). As soon as we showed a moment of weakness, and our turn seemed to be nearing its end, a crescendo of "I called next game!" would swell from the crowd.

"Are you guys talking about PewDiePie? I *love* him," a girl on the other side of the room chimed in. She told us she was one of his forty million subscribers. The whole room was now engaged in this conversation.

Isaac went back to my question. "He's really good at playing video games. It would take a lot of work to be as good as he is. So watching him is all the fun without any of the work. It's a big thing now."

Reluctantly, I asked, "What's a big thing?" After hearing my own question, I realized I didn't want to know the answer.

"eSports."

"eSports?"

"Yeah, spectator gaming where you watch gaming experts play," Isaac explained. "I'm going to a *Warcraft* tournament in a month at the Staples Center. My dad got me tickets before they sold out."

"They filled the Staples Center with people wanting to watch other people play video games?" I couldn't believe what I was hearing. I didn't know what *Warcraft* was, but I was certain it shouldn't be sharing a venue with the Los Angeles Lakers. "Do you know how many tickets they must have sold to sell out?" I pushed on.

"18,118," replied Liam.

"You're lucky. I wanted to go to that but I couldn't get the tickets in time," another student added.

I was the only person in the room shocked by the direction of this conversation. For everyone else, it seemed completely normal. Granted, not everyone was as into it as Isaac was, but for almost all of them, this was just a normal part of their lives.

As a group, we hashed out some of the particulars. We learned that the "e-athletes" sit at their gaming terminals on the "court" and the game is broadcast over the jumbotron. We discovered the winner received roughly $20,000, an amount that actually seemed a little low to us considering 18,118 people were paying good money to watch this.

When I still couldn't get over why this was a thing, Isaac said something that really stuck with me. "You like to watch people playing games, be it football, basketball, or other sports. We like to watch people playing video games. It's the same thing."

Isaac's response was eye opening for me. It was one of the first times I realized how much technology has changed the culture of young people. Traditional sports have brought different generations together for hundreds of years. On a more macro level, sports have been a pillar in most societies. Unlike tastes in art, music, film, or technology, tastes in sports have changed very little over time. The games themselves evolve at a much slower rate than virtually everything else in our society. Even within families, most children share their parents' love or distaste for them. Three or four generations of family members would gather around the television on Thanksgiving, applauding and gasping in unison. But eSports was something entirely new, a radical departure from a cultural norm. As one boy explained, "Real sports are just too slow, with too many breaks in them. They're too boring." "This is too boring" could be the mantra of this generation of "digital natives."

## The Rise of the "Digital Native Superlearners" Myth

The phrase "digital natives" was originally coined in 2001 by educational consultant and video game designer Marc Prensky. He created this term to define the newer generation of people who were born into a world in which they were surrounded by digital technology. It was intended to distinguish between the digitally fluent younger people and the older generations of antiquated Luddites who struggle to fathom the complexities of a digital world. These older people he dubbed "digital immigrants."

In an exciting time of new technological advances like high-speed Internet, laptop computers, and smartphones, Prensky predicted a dawning of a new digital era where, for the first time, the younger generation would understand the world better than the older one. Having been ensconced in digital technology from birth, digital natives would "think and process information fundamentally differently from their predecessors." So profound would their differences be, it would be as if the two groups were speaking entirely different languages, making them unable to connect in any meaningful way. This rift would require changes in long-standing relationships between young and old, primarily between parents and their children and teachers and their students.

Besides being vastly different, Prensky argued that modern technology had given "digital natives," or digiLearners, a new set of skills unimaginable to any previous generation. DigiLearners had developed the ability to multitask through their lifetime of simultaneously being on multiple digital media. Faster Internet speeds and the resulting speed at which they had information made them think faster. And their unprecedented access to limitless amounts of information gave them an insatiable thirst for knowledge.

These ideas have become the underlying theme of the educational-technology movement, whose primary goal is to "bring education into the twenty-first century" by incorporating as many screens into education as possible. Because digiLearners are so far advanced, the theory goes, schools need to profoundly adjust their methodology in order to keep up with them. The role of the teacher should be relegated to the backseat while the students, with their superior understanding of technology, should drive their own learning. Essentially, the teacher should become the facilitator of technology (or "guide on the side" as many technology advocates have titled this new role), and otherwise stay out of their students' way. The "traditional" methods of teaching are antiquated; they should be abandoned entirely. Lectures and more traditional methods of teaching are too boring for students because such methods tend to dwell on a single topic, are linear in nature, and are just too slow. Because, according to ed-tech advocates, kids today are more intellectually curious than ever before, they need to be able to jump in and out of lessons as they see fit, accessing different information along the way. Traditional classrooms only stifle this curiosity, forcing them to learn material on the teachers' schedule. Because there are so many more interesting things vying for their attention, learning needs to be made fun in order for them to be engaged.

Prensky was ahead of his time in promoting the use of computers in school. And now, fifteen years later, his claims have come to dominate educational philosophy and generate the hottest fad in teaching.

For anyone who doubts these claims that shiny new technological toys and gadgets have fundamentally improved children's skills and abilities, Prensky would say that as digital immigrants, older people simply misunderstand this generation's newfound skills. What they perceive as "kids playing a mindless game for hours on end" just illustrates their ignorance about technology. Digital immigrants' reluctance to accept change, and their pining for the "good ol' days" leaves them bitter and disgruntled. They confuse curiosity with distraction and perceive multitasking as inability to focus and new forms of communication as antisocial tendencies.

The reality is that the evidence supporting the protechnology claims about digital natives is lacking if not nonexistent. Theoretical claims are stated as facts and their evidence is mostly anecdotal. When "evidence" is provided, it is often generated by the companies selling the very technology they claim to support. However, actual evidence presented by real social scientists overwhelmingly favors the conclusion that digital technologies are bad for kids in almost every conceivable way. In 2008 a group of social scientists led by Mizuko Ito warned that we should be "wary of claims that a digital generation is overthrowing culture and knowledge as we know it and that its members are engaging in new media in ways radically different from those of older generations." And in 2009, Neil Selwyn, professor at the Institute of Education at the University of London, concluded, "The findings show that young people's engagements with digital technologies are varied and often unspectacular—in stark contrast to popular portrayals of the digital native." By 2015 Ryuta Kawashima, professor at the Institute of Development, Aging, and Cancer at Tohoku University, warned, "If rapid measures aren't taken, we may be in for serious problems. . . . It's no exaggeration to say that kids today are being controlled by smartphones, and becoming enslaved by them."

## What Real Research Tells Us

In order to understand how an outside force like digital screen technology can negatively impact our brains, we must first understand a fundamental truth about the inner workings of our brain. As Oxford neuroscientist Dr. Susan Greenfield explains in her book *Mind Change*, "The human brain adapts to its environment." This wonderfully simple and profound statement is the reason why our physical world can impact our cognitive abilities. The human brain is an extremely complex yet malleable piece of human hardware that has been the key to our evolutionary success. As our environment changes, so do our brains, evolving and adapting so they are custom fit for the needs of our day-to-day existence. Unfortunately, the new digital world is a toxic environment for the developing minds of young people. Rather than making digital natives superlearners, it has stunted their mental growth. This change is more profound and detrimental than any other generational change for three major reasons: first, the prevalence of technology in students' daily lives; second, changing attitudes toward technology; and third, the types of skills technology is starting to replace.

## The Prevalence of Technology in Kids' Lives

With an average of ten electronic connected devices per household, screen-based technologies have become inescapable. As of 2013, 90 percent of American adults have cell phones (64 percent of which are smartphones), 42 percent own tablets and computers, and 32 percent have e-readers. Undoubtedly this number will continue to rise as more "smart" screen-based electronic devices are created. The demand for even more access to smart devices has led to the creation of devices like the Apple Watch (an iPod Touch with a wristband) and Google Glass (eyeglasses that have a transparent computer screen as a lens). Both of these devices provide their users the ability to check e-mails, send texts, shop, and even read the newspaper at any time by literally attaching screens to the users' arms and faces. Now, connected screens even adorn our refrigerators, in case your trip to get a snack has left you feeling too disconnected.

Even with all these screens already in our lives, the number of devices in the typical house is expected to double in the next five years. Running out of places to attach new products to their adult users, electronics producers are looking to exploit new markets by appealing to younger and younger demographics. Today, 56 percent of six-year-olds have their own cell phones and 80 percent of twelve- to eighteen-year-olds have them. Over half these phones are considered "smartphones," which possess all the functionality of a computer with the portability of a standard cell phone. Screen time is also an issue for our youngest consumers: 29 percent of infants spend ninety minutes or more a day on screens, toddlers are averaging two hours a day, and by age four that number doubles. These numbers will continue to rise at alarming rates as the number of apps and devices marketed to children also increases.

The increasing portability and affordability of screen-based electronic devices have made them much more pervasive in the modern environment. There are now powerful technological devices with seemingly limitless potential in our pockets, on our wrists, directly in front of our eyes, and even in front of our infants and toddlers while they sleep and go to the bathroom. The sheer availability of screen technology has obviously led to unprecedented exposure to it. According to one of the most comprehensive surveys done on digital natives' technology use by the nonprofit group Common Sense, teens are averaging almost nine hours a day on entertainment media. This includes self-amusement activities like participating in social media, watching television shows and movies, listening to music, and playing games. Tweens (eight- to twelve-year-olds) aren't too far behind, averaging almost six hours a day. This time does not include that spent on screens during school, for homework, or even time spent talking on the phone or texting.

Nine hours is a significant portion of the digital native's day, to say the least. Teens are spending more time on screens than they spend on any other activity in their lives. It's more than the eight hours a day they spend at school, and more than the roughly seven hours they spend sleeping. After spending this amount of time on their phones and other devices, there is virtually no time left for more traditional childhood activities such as studying, playing outside, meaningful face-to-face conversations with friends or family, and hobbies. This is one of the greatest dangers of overusing technology—the opportunity costs. It's the major life functions and experiences that are being given up for technology that are having the greatest impact on child development. If modern technology could be used in moderation, that would be one thing. But the most appealing technology on the market today is designed with the intent of being addictive, to increase the duration and frequency of its use. These technologies are dominating every facet of a young person's life. As if the nine hours a day weren't enough, schools are now adopting initiatives that encourage and in some cases require kids to be on screens for much of their school day as well. With kids of the near future spending nearly fifteen hours or more on screens, electronic media is no longer having an impact on their environment. It is their environment.

And it's not as though teens are using their phones for learning, creating, or other productive pursuits. I can honestly say I've never taken a phone from a kid who was in the middle of exploring a cyber art museum. I've never had a parent complain that she walked in on her son having a late-night FaceTime session with a group of school children in Nigeria. Pretty much all I ever see kids do on their technology is text and Snapchat friends, play games, take pictures of themselves, check Instagram for likes, watch silly videos, and play more games. According to the Common Sense census, 78 percent of the time teens spend on their electronic devices is devoted to "passive" and "interactive" consumption. These categories include "watching, listening, reading, and playing with media content created by someone else." Young people only spend about 3 percent of their digital time in "content creation." This would include things like real writing, taking a creative photograph, composing a song, coding, or any of the other mentally stimulating activities technology has to offer. This lack of creation and overabundance of passive entertainment is the major reason for digital natives' shortcomings.

If in 1990 a mother claimed her kid was smarter because he was spending most of his waking day on his Game Boy playing *Mario Bros.*, people would have thought she was insane. But this is essentially the claim of the protechnology camp today. Any logical person could conclude that the nine hours a day children are spending passively entertaining themselves has done more to digitally lobotomize them than it has to make them more intelligent.

Something happened in 2001 that made it harder for young people to find the more productive side of technology. The same year that Prensky published his paper "Digital Natives, Digital Immigrants" an equally influential paper was published by Dr. John Hopson. In his piece entitled "Behavioral Game Design," the PhD in behavioral and brain sciences from Duke University outlined for the gaming industry how to keep their players playing longer. Hopson applied the same methodology for the video game industry that was used by behavioral scientists to create addiction in lab rats. By teaching game designers how to generate a similar response in their users, he created a framework for addiction. Today, these elements of addictive gaming can be found in virtually every gaming app on the market. They include tricks such as incentivizing playing through some type of reward (unlocking new levels, receiving tokens, obtaining a coveted object, etc.), variable-ratio rewards (randomized incentives spread throughout the game such as an extra life or a hidden sack of treasure), increasing this rate of reward over time (getting the player hooked by frequent rewards at first, but extending the intervals of reward to longer periods as the player continues), and punishment for avoidance (taking away rewards for not playing every day, or breaking "streaks").

These are the very same tactics casinos use to ensnare gamblers. Gambling addiction is so common that many casinos are required to have phone numbers for gambling addiction hotlines posted on all their ATM machines. A video game designer for *The Settlers Online*, Teut Weidemann, let it slip to a group of reporters at a gaming convention in 2010 that in order to continue "monetizing all the weakness of people," his games had to "bring them in and keep them addicted and make them keep playing." Similarly, these tactics have been employed so effectively in games like *World of Warcraft* that there are countless stories of extreme addictive behavior associated with their use. In 2014 a young California couple was arrested for child neglect after they kept the children in their care locked in their rooms while they went on a three-year *WoW* bender.

A 2009 study done by Iowa State psychologist Douglas Gentile found that 12 percent of boy and 8 percent of girl gamers demonstrated "pathological patterns of play" and all the signs of addiction. But these numbers don't properly illustrate the severity of this disorder. Teachers regularly witness the impact tech addiction of all kinds has on the behavior of young people. It can manifest itself in many different ways. The most obvious is their inability to put down their phones. For several weeks I kept a tally of the number of students who walk the hallways clutching their cell phones in their hands rather than putting them in their pockets or purses. The result? Seventy-five percent do this. When I asked several students why this was, they replied that they were afraid they'd miss a text or other type of message—a classic "fear of missing out."

Other signs of addiction include exhaustion during the day. The week after winter break one year, I had a female student who couldn't stay awake during class. Despite waking her up repeatedly, she just couldn't keep her eyes open for more than a few minutes. Worried that something was seriously wrong, I pulled her aside after class to see what was going on. She confided in me that her parents had gotten her an Xbox for Christmas. They would only let her play for three hours a day, so she was setting her alarm for 2:00 AM and playing while they were asleep. She would stop when they woke up and get back in bed just in time for them to "get her up" for school. She had been doing this for weeks.

Overuse of technology can have more serious effects. One year, I had a senior who was doing poorly in many of his classes. As a county policy, he needed to pass my economics class in order to graduate. Not wanting to be the reason he didn't get a high school diploma, I made a deal with him: if he did well on his final paper and passed his final exam, I would bump his grade up to the lowest possible, D-. Considering how low his F was, this was a generous offer. The essay was to be an analysis of the recession we were experiencing at the time. In the directions I provided him, he was to work in concepts like inflation, stagflation, unemployment, underemployment, GDP, and complementary goods, all major concepts of our course. The day came when the assignment was due. I was eager to see what he had done and hopeful that he cared enough about his education to put some effort into this assignment. But what he turned in to me was shockingly bad. It was a printout of a conversation about a 9/11 conspiracy theory in an online forum. Not only was it completely off topic and lacking any type of actual evidence, it was 100 percent plagiarized. He had simply copied a crazy person's accusations he had found on Reddit and pasted it into a Word document.

The parent conference we had is something that I'll never forget. It was led by an older assistant principal (very much a digital immigrant). She was an old-school, no-nonsense administrator who wasn't afraid to speak plainly. She could be loving and supportive when her students did something positive, or she could be ninety-five pounds of pure terror when they did something negative. Not one to mince words, immediately after introductions, she looked the student square in the eyes and asked, "Son, are you on drugs?"

The student's dad immediately jumped to his son's defense, "No, no. We can assure you that he's not on drugs."

"It sure looks like drugs to me," she insisted. Holding his transcript in the air, she continued, "He went from being a B student last year to suddenly failing all of his classes. He's got huge bags under his bloodshot eyes, he's disheveled and doesn't look like he's showered in a few days. So are you sure he's not on drugs? Sometimes kids are good at hiding that from their parents."

Caught off guard, his dad went back on the defensive. "It's just that he's been playing this anime video game for twelve hours a day over the last several months—he's not doing drugs."

The administrator's jaw dropped. Shocked, she turned back to the student. "Let me understand this. You're failing out of school because of a video game? What is wrong with you?"

His addiction had become so strong that he was willing to forfeit everything he had worked for just for this game. He was risking forfeiting a high school diploma for it. This is the hallmark of addiction. And he's not alone. Former commissioner of the Federal Communications Commission Deborah Taylor Tate drew the ire of the tech industry in 2008 when she claimed obsessive gaming was "one of the top reasons for college dropouts in the US." Kids who struggle to control their compulsion in high school, when they're under parental supervision, have little chance when given the freedom they experience in college.

But Tate's remarks were largely dismissed because she lacked any hard numbers. And her critics certainly had a valid point— she didn't have any objective research to back up her claims. She was basing her statement on firsthand accounts from students she had encountered while going from college to college. But that is also a part of the problem. Unfortunately, there have been few studies into the effects of Internet use disorder (also known as Internet addiction disorder) in the United States because, unlike most of the developed world, it is not recognized in the United States as an actual disorder. Countries like Australia, China, India, Italy, Japan, Korea, and Japan not only recognize it but also have established numerous clinics for treating it. The *Diagnostic and Statistical Manual of Mental Disorders* (*DSM*), the primary reference for health care providers for diagnosing mental health conditions, wasn't updated to include gaming addiction under the title Internet use gaming disorder until 2013—and even then, it was with the caveat of "for further study."

## Changing Attitudes Toward Technology

As a society, our attitudes toward the use of technology have dramatically changed in the past decade. Not long ago, a child spending too much time on his or her computer or gaming system was viewed by many parents as problematic. Parents would put limits on their children's technology use and encourage them to go outside and play, make friends, or only game once their homework was finished. Today, however, there is an increasingly common attitude of acceptance. With an air of resignation, parents repeat over and over the mantra "This is just how kids are today."

This acceptance is the result of a carefully choreographed series of assumptions perpetuated by the protechnology movement. The first assumption is that "regulating technology use is just too difficult." When technology was big and expensive, TVs and desktop computers for example, there may have only been one or two stationary devices in the home. Monitoring was easy for parents. It was impossible to hide a TV or desktop computer, and it was easy to box it up when kids abused it (my parents would simply hide the controllers, leaving the gaming system behind to make it more torturous). Today, because of the sheer volume of devices, parents face a daunting task in putting limits on screen time. Coupled with the increasing portability of devices, keeping tabs on a child's tech use is much more challenging than it was just a few years ago. How can a parent possibly keep tabs on his or her child's phone or Apple Watch use?

For a teacher, the task of regulating modern technology is every bit as daunting. With every student coming into the classroom with one to three devices of his or her own, a teacher could be stuck trying to oversee the proper use of fifty or more devices. Students can hide them in their laps, pockets, or in plain sight. Once, after confiscating a student's smartphone, and then their Apple Watch ten minutes later, I found him playing a game he had downloaded to his graphing calculator—his last resort for digital amusement. This was just one of my thirty students. For many teachers, this can quickly turn into an impossible battle to win.

The next assumption is that this battle isn't worth fighting because tech addiction is not a bad thing. When looking at any claim about the addictive nature of gaming or technology, the most common response is, "So what?" Unlike alcohol, tobacco, or drug addiction, there isn't a visible impact on one's health (although there's a strong correlation to levels of obesity and time spent gaming). Unlike gambling addiction, there isn't the immediate negative consequence of losing all one's hard-earned money. And unlike pornography or sex addiction, there isn't the seedy immorality associated with gaming or social media, nor is there the immediate damage to relationships. The common rationale is that screen activities bring people pleasure without any of these negative side effects, so why worry about it?

This is the false assumption that allows tech developers like Teut Weidemann to openly admit to intentionally designing their products to be addictive, and then market them to children. This is something that big tobacco had to go to great lengths to conceal from the public. Video game designers, though, brag about it to journalists at gaming conventions.

If parents were forced to choose a gambling, drug, porn, or *World of Warcraft* addiction for their children, I would guess most parents would choose *WoW*. But not being as harmful as gambling, drugs, or porn is far from a glowing endorsement. As I will cover throughout the book, there are many negative side effects associated with technology addiction, most notably increased depression, anxiety, withdrawal, diminished focus, and diminished cognitive function.

As many researchers are beginning to find out, it's not just gaming that is addictive. A teacher from Kansas whom I interviewed was having an issue with a student who was not completing any of her work. Rather than participate in class, she would spend the entire period on her phone with social media. Because of this, she was failing for the year. It's worth noting that this teacher had heavily incorporated technology into all his lessons. Every class, each student was given a laptop or iPad to complete his or her work. However, that didn't lessen this young woman's compulsions because it wasn't the act of using electronics to which she was addicted. She was specifically addicted to social media. She viewed her class as something that got in the way of that.

The problem worsened as the student grew increasingly defiant of the teacher. When the teacher had finally had enough, he threatened to take the phone away unless she put it away herself. She didn't. That's when things got ugly. As the teacher reached for her phone, the girl reacted violently. She lashed out at the teacher, screaming, "You can't touch me or my phone! Get off of me!"

After the incident, the teacher debriefed with the student's guidance counselor. Between her work with digital natives at school and in her private practice as a therapist, this counselor had become very familiar with tech addiction, and so this student's violent reaction didn't come as a surprise to her. She told this teacher that when dealing with many digital natives, you have to treat them like addicts. Just as you would expect a violent reaction if you were to take a bottle from an alcoholic, or a needle from a heroin addict, you can expect the same type of reaction from a student with a tech addiction. She's counseled several parents who had to call the police when their children turned violent when they attempted to take away cell phones, limit social media use, or delete their gaming accounts. She's also dealt with a client who failed out of college twice because of his technology addiction. She's had to counsel a twenty-five-year-old who barely graduated high school, never went on to college, and had spent the last seven years in his parents' basement, unemployed and completely supported by his parents. How could anyone say this addiction is "harmless"?

Once one believes that fighting the tech trend is futile, and that kids spending all day on their devices is not harmful, they've been primed for the third assumption—spending all day on technology is actually a good thing.

In one of his latest books, *Don't Bother Me Mom–I'm Learning!*, Marc Prensky urges parents and educators to essentially leave their kids alone while they're playing on their devices and let them use their technology however they want. He argues that digital natives are playing video games to learn and improve their cognitive abilities.

In 2014 Microsoft researcher and ed-tech advocate danah boyd wrote a piece for *Time* magazine entitled "Let Kids Run Wild Online" in which she advocates for children to be allowed to explore the Internet and social media free from parental constraints. She labeled parents' monitoring of their children's online activities as "helicopter parenting" and likens it to parents who won't let their kids go outside or go to a playground unmonitored. That's not a reasonable comparison. Imagine a playground where there were stashes of porn under the slide and the swings, and a significant number of the other "children" were actually forty-year-old sexual predators in disguise. Any reasonable parent would either not allow the child to go or would watch that child's every move while she was there.

The problem with these assumptions is that, as we already know, children aren't spending their time on technology in productive ways. They're spending it gaming, listening to music, talking with friends, and watching TV or movies. Kids aren't using their devices to learn, because learning is hard work. Kids aren't spending their time realizing the potential technology has to offer because they can't stop using it for recreational leisure activities. To say that this generation of children is profoundly different than any generation of children to come before them is unsupported by any real evidence. It also defies logic to assume that laptops and iPads have somehow transformed children's predisposition to play into a new desire to learn and work hard. As neuroscientist Susan Greenfield points out, "If you think about this claim, it's based on the assumption that the only thing keeping flocks of young people from filling up libraries for the purpose of voluntarily researching and learning on their own was they didn't have a ride." But hand a kid an iPod, and now it's reasonable to expect him or her to use it primarily for educational purposes? Only someone who has spent almost no time with children could possibly believe that.

This propensity toward play is reflected in the ed-tech movement. For the most part, teachers aren't using advanced technology like data analytics, spreadsheets, coding, or other technology common in the modern workplace. Rather, schools are trying to use "edutainment" and educational gaming in an attempt to sneak learning into the digital playground. Teachers resort to lessons like "Create a Facebook profile for Thomas Jefferson" or "Tweet your opinion about Holden Caulfield." During a recent in-service for educators on the topic of increasing technology use in school, one teacher told me about how he was using the video game Minecraft in his elementary class. He admitted that he had been having his kids play the game for weeks. However, it was only after he was asked to give this presentation that he had his students "hypothesize" about the lessons they had learned while playing the game. Even though the children were very much "engaged" in the game, the educational purpose of his exercise was clearly only an afterthought.

This is the reality created by idealistic claims of those who urge educators to "meet digital natives where they are." Teachers aren't stepping up into digital natives' alleged advanced digital world. They're having to dumb down everything in order to step down into kids' simplified digital existence. Rather than deal with the issues created by technology addiction, schools are trying to trick digital natives into learning by sneaking small, palatable doses of education into their games and social media.

In the most traditional method of education, play is incrementally replaced with rigor as students grow older. Kindergarten is mostly spent playing and socializing, with basic educational concepts woven throughout. However, grade by grade, play time is taken away. By senior year in high school, play has mostly been replaced with rigor and challenging educational work. The idea behind this is that as the child matures into adulthood, education would transition to prepare students for the demands of the real world.

The sentiment behind danah boyd's remarks on allowing children a space to be free and play is certainly valid. With the passage of No Child Left Behind in 2001, school systems began switching to more standards-based learning supported by constant data collection. Play in early elementary school was largely replaced with work. Recess was replaced with lessons on difficult math concepts or advanced reading techniques. Parents lamented the loss of their children's childhood. No longer did children know how to play.

Ed-tech firms responded quickly. Parents and school systems could have their cake and eat it, too. By providing children with educational gaming software, students could play while learning. The educational gaming industry thrived despite lacking any evidence as to its effectiveness. The notion spread. If learning could be made fun for younger students, why not make it fun for older students?

Even if the claims of educational gaming app developers were true (which, as I will address later in the book, they are not) and gaming apps were as effective as traditional methods of educational instruction, there is still a fundamental flaw in the logic. If a student's entire educational career, from kindergarten to senior year of high school or even college, were spent primarily using such edutainment apps, with "fun" being the focus rather than "rigor," at what point would a maturing adult learn how to put aside play in order to work? Would this burden become the prerogative of the workforce? Would companies of the future have to adapt by creating gaming apps to fill out paperwork, obstacle courses to complete physical tasks, and number munchers to complete calculations?

Digital natives' addiction to technology is so strong, however, that many digital natives refuse to grow up and ever even enter the workforce. The stereotypes of digital natives stuck in their parents' basements glued to their screens rather than getting a job is becoming more of a reality. Millennials are delaying becoming functioning members of society for longer than any group in recorded history. Today, "Living at home with their parents" is the most common living arrangement for people age eighteen to thirty-four. This is the first time since 1880 (when this data was first collected) that more of this demographic lived with their parents than a spouse or partner or on their own. A recent collaborative study by economists at Princeton, the University of Rochester, and the University of Chicago found a strong correlation between unemployment and underemployment of millennial men and addiction to gaming. The number of twenty-two- to thirty-year-old men with less than a bachelor's degree who reported not working at all in the last year has more than doubled since 2000, 9.5 percent to 22 percent. In that same time frame, the time this demographic has spent on technology has gone from 3.4 hours a day to 8.6 hours. As a twenty-two-year-old man, still living with his parents, explained, "When I play a game, I know if I have a few hours I will be rewarded. . . . With a job, it's always been up in the air with the amount of work I put in and the reward."

Given these realities, it's baffling that society's attitude toward kids and technology has somehow become a collective resigned shrug.

## Technology Is Replacing Important Fundamental Skills

The third reason why today's technology use at home and school is so much more damaging than it was even a decade ago is that for the first time in history modern advancements are prompting the wholesale abandonment of most of our prized brain functions. We're willingly outsourcing them to our technology. I see this in my classroom every day.

Socrates's fear that the development of writing would inhibit man's ability to memorize information seems absurd today, but not because it isn't true. The reality is his prediction was more than likely correct. Consider one of Socrates's points of reference, the famed Greek storyteller Homer. Despite being illiterate (as were most Greeks of that era), Homer is credited with composing and memorizing the entire works of the *Iliad* and the *Odyssey*. These works alone would be the equivalent of a modern thousand-page book. Memorizing them would be a seemingly impossible feat today. But in a world without writing, memorization would have been an important part of one's environment, and the brain would adapt by allotting additional resources to make it more successful at that task.

But as Marc Prensky points out in his 2009 article "*H. Sapiens Digital*: From Digital Immigrants and Digital Natives to Digital Wisdom," "Every enhancement comes with a trade-off." Mr. Prensky discusses a future in which the shortcomings of our brains will be "enhanced" in ways that will make us all wiser. For example, although spell-checker will lessen our ability to properly spell words on our own, language translation software and social media will allow us to universally share ideas instantaneously. Although calculators will eliminate our ability to solve basic mathematical problems on our own, computer-based algorithms will make complex computations that have all sorts of real-life applications. Computer-based simulations will eliminate our need for real-life experiences. Because of the increased prevalence of technology, almost all our cognitive abilities will somehow be digitally "enhanced." This goes way beyond bits of our memory being traded in for the development of writing. In Mr. Prensky's opinion, these mental trade-offs for improved technological support are so profound, he believes modern humans are deserving of a new trinomial nomenclature: *Homo sapiens digital.*

Prensky supports this with a quote he heard from a teen, "If I lose my cellphone, I lose half my brain." Call me pessimistic, but rather than finding this statement to be an affirmation of a marvelous future in which cell phones make us smarter by giving us cognitive abilities unimaginable before the turn of the millennium, I find this statement deeply disturbing. Teens like this aren't using the advantages technology offers as a new starting point from which to reach new mental heights. Instead, they are dependent on their technology to do all their thinking for them—more mental crutches than mental amplifiers. For example, without a basic understanding of fundamental mathematical concepts, who will generate the complex mathematical algorithms of the future? The problem with almost all learning is that it is sequential. This is what makes learning so challenging, and so resistant to shortcuts. Just as infants don't learn how to run before they can walk, a person can't make sense of complex mathematical equations without an understanding of how to add, subtract, multiply, and divide. To say this ability is no longer relevant in an age of complex calculators is to misunderstand the importance of knowledge.

Over the last several years I've sought to quantify the pervasiveness of this misconception by asking students the following question: Should you be required to learn information you can look up? Not surprisingly, as more and more students have access to smartphones with apps like Google, which can instantly conjure up almost any information imaginable, the number of students who believe they should not be required to learn searchable information has sky-rocketed from roughly 1 percent to 35 percent in just five years.

The problems with this mentality are many. First, Google can give us information about nearly any imaginable question, so why should digital natives bother to learn anything? Second, you have the more practical question, What if you lose your smartphone; then what do you do? Having outsourced all their basic cognitive functions to their electronic devices, what are teens to do without their mental crutches? But this is more of a pragmatic question, to which tech companies will respond by surrounding us with more devices. The deeper problem is that looking up an answer mistakes "answering a question" with "learning." Because Liam Googled how many seats are in the Staples Center doesn't mean he now knows it. If you were to ask him the same question a day or even an hour later, he likely wouldn't remember the correct answer. Google gave him the ability to answer a knowledge-based question without him internalizing the knowledge.

This is especially true if Liam didn't have any context for the Google-generated answer. If he had never heard of the Staples Center, never seen a picture of it, didn't know who the Lakers were, or had never been to a basketball arena, then to him 18,118 would be just a random number without significance. The brain tends to discard random facts like this because it can't make sense of them, and therefore decides they must be unimportant.

This is the true importance of genuine internalized knowledge. Think of the brain as a water pipe and a bit of information as a speck of dirt. When the pipe is empty and free of particulates, that speck of dirt passes easily through the pipe. In this analogy, the bit of information had nothing to grasp on to, so it was discarded rather than retained. However, as more and more specks of dirt start to accumulate, it becomes easier and easier for new specks of dirt to find somewhere to hold on to, making it harder and harder for new specks of dirt to pass through the pipe without being retained.

Real learning begets new learning. Learning becomes easier as we increase our knowledge base because we can connect new information to things we already know. And this is why teachers are so important. Simply telling someone a fact, or telling someone to look up a bit of information, rarely makes that person learn it. Good teachers know how to make this information stick in their students' minds. Having what may seem like a meaningless fact stuck in their minds today helps students make sense of important concepts later on. By maintaining the idea that knowledge is unimportant, digital natives have no framework to make sense of any new information. Their pipes are squeaky clean, void of any true knowledge. New information and concepts have nowhere to stick. Although able to generate answers, without any context they will struggle to understand and internalize what they're saying. Simply handing them a tool that gives them correct answers doesn't make them wiser. A student like Liam can become a "know-it-all" without really knowing anything at all.

This is what digital immigrants will never understand about digital natives. Technology is most effective when being used by someone with a strong fundamental base of knowledge in the area that a particular technology works to enhance. To a digital immigrant, someone who is starting off with a base of unenhanced, self-developed knowledge, technology can act like a springboard, expanding the potential of existing abilities. However, for the digital native, who has a reduced baseline of fundamental knowledge from a lifelong dependency on technology, technology offers little potential for enhancement because there is nothing to enhance in the first place. Even if technology can improve existing cognitive abilities a thousandfold, one thousand times zero is still zero. (I did that without a calculator.)

Because it is increasingly common for today's kids to outsource knowledge acquisition, their base of knowledge is a foundation built on sand. Never having to learn anything because they have technology to do their learning and thinking for them has left them incapable of using technology to improve upon existing bases of knowledge.

To put this in context, when the elderly discover social media, they apply their real-life understanding of social interaction to it. My eighty-five-year-old grandfather just recently got on Facebook. He did this not to replace existing social interactions but to enhance these interactions. He continues to write, call, and see his family and friends with the same regularity as he did before. However, Facebook allows him a chance to increase the frequency of interaction in a way that is more dynamic and timely than writing a letter, but not as fulfilling as actually being with the people with whom he chooses to interact. He brings his lifetime of knowledge acquired through face-to-face interactions to every type of social interaction he has. Whether it's a phone call, text, e-mail, or Facebook post, he can accurately predict what type of remark will elicit what type of response from the recipient. He can also differentiate what setting is appropriate for a formal or informal tone. Despite being new to the technology, he picked up the nuances of Facebook instantly. "How'd you figure out how to like that post of mine?" I asked him. "I'm old, not dumb," he reminded me. "I can figure out what a thumbs-up means."

Unlike my grandfather, kids today do most of their socializing through social media, and they haven't spent a lifetime having face-to-face conversations. They've replaced body language with emoticons. They can choose what comments to acknowledge and which ones to ignore, something much more difficult in actual conversation. They don't have to see others' reactions when they say something offensive, or hear the deafening silence after a joke falls flat. By choosing to primarily interact with people through this inferior yet more convenient form of communication, their social abilities have diminished. Social media doesn't enhance today's kids' existing social interactions because they lack the skills and knowledge that are supposedly being enhanced.

If tech advocates' predictions of an enhanced technological future were true, you would expect to see technology improve existing forms of communication. Yet which social media has improved upon a face-to-face conversation? E-mails? They're just faster forms of letters—lacking any emotion or body language. Texts? They're just short e-mails. Tweets and posts? They're texts made to a broad audience who may or may not choose to look at them. How about the latest social media fad, Snaps? They're just texts of pictures, typically selfies (photographs in which the subjects and photographers are the same, and normally limited to just their faces) where they may or may not have added a cartoon picture of a funny hat.

For years I have been working with your kids for eight hours every day. It is clear to me that as they retreat more and more into a digital world, their communication skills have worsened. When confronted with that idea, I hear ed-tech supporters predictably reply, "You just don't understand." Teachers may fumble through modern-day vernacular, like saying "hashtag" when they meant to say "tag," but that doesn't create a rift so deep between students and teachers that we can no longer possibly communicate with each other. These claims by ed-tech advocates are so exaggerated they're laughable.

Educators should continue to instill in their students important and fundamental skills and knowledge. We need to build the critical skills that have always been central to the human experience. In order to do this, schools need to lessen technology use rather than increase it. What I have attempted to show, though, is that we are in an alarming new world in education. More and more that world is disconnected from the real world and plugged in to a virtual one. However, children need to be learning how to use their brains, not their iPads. When we get the focus back on using their brains, their ability to think critically will return as well. I take a closer look at this in the next chapter.

## TAKEAWAYS

- Question what you hear from tech proponents. Be wary of unsubstantiated claims. Many of these "tech gurus" and edutainment creators are making assertions without solid research or evidence to back them up. And when they do cite "research," the studies are not peer reviewed and typically claim technology improved areas of child behavior that are notoriously difficult to measure, such as student engagement. Even then, look to see who funded the studies. Much of the research being done is paid for by the people who are selling technology and who stand to profit when school systems invest in their products.
- Look to unbiased and objective experts. There is a lot of great peer-reviewed work by legitimate scientists from a wide variety of fields on the topic of technology's impact on the human mind and behavior. Although I reference their findings throughout this book, their full works are worth a read. They include Oxford neuroscientist Susan Greenfield, MIT technology and social professor Sherry Turkle, Stanford sociologist Clifford Nass, George Washington neurologist Richard Cytowic, and psychologist Richard Freed, just to name a few.

- Get real. For a child, a screen is, at its core, a toy. Regardless of which generation children are born into, expecting them to ignore the entertainment side of technology to realize its full educational benefits is unrealistic. And even educational games do more to addict kids to gaming than improve their skills and knowledge.
- Voice your concerns to your children's schools. Many school systems are more responsive to parents than teachers. Be an advocate for in-person human connections and time away from screens.

3

# Reclaiming Your Child's Ability to Think

*I fear the day technology will surpass our human interaction.*
*The world will have a generation of idiots.*
—Albert Einstein might have said this. But he probably didn't
(regardless of what the Internet says)

ONE OF THE HOTTEST TOPICS in education right now (besides how to incorporate more technology into classrooms) is how to improve students' critical thinking. Over the last decade, teachers across the country have bemoaned a marked decline in their students' ability to think for themselves. Former staple activities such as creative writing prompts, formulating opinions, developing arguments, or answering open-ended questions are becoming increasingly challenging for students. Many teachers have simply abandoned them because their efforts to generate deeper levels of thought seem futile. Everyone is trying to figure out ways to improve critical thinking in students, but no one seems to be trying to figure out why these skills were lost in the first place.

## Research? What's Research?

You might find this hard to believe, but there's never been a law in the United States about abortion. There are no primary sources on the Bush administration, or even the Crusades for that matter. And the Internet doesn't contain one iota of information about the Laffer curve. These are the things I am told when I ask my students to do research.

Doing research for school used to be difficult. It required trips to the library, indexes, card catalogs, and the Dewey Decimal System for God's sake. You had to think about what potential sources of information might be before you started looking for resources. It was like detective work, and finding relevant supporting evidence was like searching for clues. To be successful, you had to explore different angles and be creative in discovering new places to look for information.

Today, technology has made research infinitely easier. Search engines like Google can turn up almost any information no matter how random or obscure. I once downloaded a digital copy of the artist Prince's high school yearbook to settle a lunchtime debate between two teachers. It took five minutes to find. Gone are the days of bouncing from library to library looking for the right book and scrolling through microfiche. There's an endless world of information that's just a click away. There's information on any conceivable topic accessible by any connected device. Modern technology seems to have solved every problem a would-be researcher might encounter.

But you wouldn't know this by talking to many digital natives today. Giving modern students any question that doesn't include in the instructions explicit directions on where to find the answer is only done by a teacher who is new and inexperienced or a glutton for punishment. In social studies, rather than traditional research papers, we've moved to DBQs, or document-based questions, which are simply research papers where the teacher has done all the research for the students, and gives it to them in a neatly organized packet.

Not me, however; I'm one of those aforementioned gluttons for punishment. For my history students' end-of-the-year project, I have them do an old-fashioned research paper. They can pick whatever topic they like as long as it falls within the curriculum. They spend two weeks learning about the different types of evidence, developing outlines, and planning strategies for research. The librarian even gives students a tutorial on how to use the dozens of research databases to which they have access. I go so far as to give the kids a list of useful phrases to type into a search engine to help yield better results—phrases like "firsthand account" or "eyewitness testimony."

But every year, this same assignment seems to get harder for students. Paradoxically, the easier technology has made research, the more kids struggle with it. "There are no primary sources on my topic," one ninth-grade honors student sobbed, "and I've spent the last three days looking."

I went through the checklist created for the students. Did he look at the database of primary sources? He had. Did he use the suggested phrases in the search engine? Yep, did that. Had he examined the sources in the bibliography of a secondary source? He had done that as well.

"I can't do this project. It's too hard," he said while throwing his hands in the air in an act of surrender. Finally, I asked what his topic was. "The Bible," he replied.

"The Bible?" I couldn't believe what I was hearing. "You can't find a primary source on the topic of the Bible?" I asked to make sure I had heard him correctly.

"Yeah, there's nothing," he answered.

"The Bible *is* a primary source on the Bible," I tried to explain.

"Oh, OK, but you said the primary source had to be in print," he replied. I hadn't said that. The directions said one of their sources needed to be in print and one of their sources needed to be a primary source. But I could tell he was already too frustrated for me to explain this nuance.

"Why not just use a print version of the Bible?" I asked.

"Are there any?" he wondered.

"Are there any print versions of the Bible?" I like to repeat questions like this back to the student. Sometimes hearing their own words makes them realize the answer.

"Yeah," he answered.

My trick hadn't worked. "It's literally the most printed book in the world," I replied.

"Where can I find one?" he demanded.

"Pretty much anywhere; this library has five," I said in an attempt to lift his spirits. He still seemed unsatisfied, so I pointed to one of his classmates who was sitting at the table adjacent to his. "Shelia has one. Ask her if you can borrow it." Finally, the problem had been solved. The story of this young man embodies the struggles many of his peers have solving basic problems, even with technology designed to do it for them. He's not alone.

## More Technology, Less Ability to Problem-Solve

The 2012 Program for the International Assessment of Adult Competencies, a test administered by the Organization for Economic Co-operation and Development (OECD) to compare the cognitive and workplace skills of people across much of the world in the areas of literacy, math, and problem-solving skills, did not paint a flattering picture of American education. Of the twenty-four countries that participated in the test, the United States ranked second to last in math, and dead last in both literacy and problem-solving skills.

A portion of the test was designed to measure the problem-solving skills of sixteen- to twenty-five-year-olds by "using digital technology, communication tools and networks to acquire and evaluate information, communicate with others and perform practical tasks." According to the OECD, this wasn't a simple "measurement of computer literacy, but rather of the cognitive skills required in the information age—an age in which the accessibility of boundless information has made it essential for people to be able to decide what information they need, to evaluate it critically, and to use it to solve problems."

The questions on the test were designed around real-world problems. One set of questions had the participants evaluate the websites generated by a generic search engine's results for the phrase "job search." The user was asked to evaluate each link and bookmark the sites that did not require a paid subscription. As with most real-world websites, many of the simulated ones did not immediately advertise that they required a paid subscription and only made it apparent after the user began to interact with the website. However, this varied from site to site, just as it would in the real world.

Having been in the classroom as long as I have, I can imagine that many of my students struggled on this test because there was no concrete way to arrive at the answers. Problem-solving methodology had to adapt from website to website. The instructions didn't explicitly tell the test takers how to arrive at the correct answer.

This flies directly in the face of many claims today, such as those by Marc Prensky, that a digital native requires less formalized, step-by-step instruction. If a lifetime of exposure to technology made students more intuitive (as the digital native claim goes), then our digital natives should have thrived on this portion of the test. Of all the countries that participated, the United States invested the highest percentage of GDP into information and communication technology from 1990 to 1995 and the second highest from 1995 to 2002. The sixteen- to twenty-five-year-olds of Finland (which invested less than half of the GDP percentage the United States did in technology from the years 1990–1995) ranked second in their ability to solve this technology-based problem.

The failure of the US education system's ability to develop problem-solvers is not from a lack of technology use. Simply using technology doesn't make one "tech savvy." Tech savviness begins with one's ability to problem-solve. And it's only when this ability to think creatively, critically, and logically is paired with an even more basic understanding of technology that one becomes "tech savvy." Without problem-solving skills, one might be able to learn how to operate a specific program or piece of hardware but struggle to transfer that understanding to a closely related program. Recently, I heard an art teacher complain that when his class switched from editing photos in Microsoft Paint to Adobe Photoshop, most students couldn't figure out that the pencil icon in Paint did the same thing as the pen icon in Photoshop. To anyone with basic problem-solving skills, this would have been immediately clear.

Paradoxically, this inability to use technology to its full potential may be a result of kids' excessive technology use at an early age. As neuroscientist Susan Greenfield explains, "Video games are replacing children's imaginations." Rather than playing in a world constructed by their own minds, worlds in which they're cops and robbers, doctors and patients, princesses and aliens, or whatever they'd like to be, kids today play mostly in worlds created by others in the form of video games or other media content. Make-believe is a mentally rigorous exercise that helps a child develop critical thought and creativity. But kids are systematically replacing this important form of play with games created by the minds of others.

A 2005 multinational study illustrated how influential media consumption can be on a child's imagination. It followed eight- to ten-year-olds' media consumption and then tracked their "daydreams." They found that most children's imagination deviates very little from the presented story line. Their daydreams are almost a verbatim retelling of the story they just watched or the game they just played. Because technological forms of play typically require no imagination, the portions of the brain children used for it become underutilized and, as a result, do not fully develop. It's these forms of early play that can create imagination, which can lead to creativity later in life. Lacking in creative thinking can likely lead to difficulty with outside-the-box thinking, which is essential for problem-solving.

Children with little imagination grow up to be adults with poor problem-solving skills. Recently, I participated in a forum led by local business leaders. This presentation was on the necessary skills needed in the modern workplace, and what educators can do to help. Despite the fact the presenters represented a variety of different companies, they all had a similar message: digital natives cannot think for themselves and lack basic problem-solving skills. Employer after employer lamented that they have to instruct millennials on every little facet of a job because they have difficulty inferring how to proceed in solving problems.

One of the most striking stories I heard that day came from the owner of a video production company. The employer was telling the crowd of teachers how he usually assigns the tedious job of recording conferences to his new hires, who are typically fresh out of college, to help them get their feet wet. Assuming all his new staff have backgrounds in media arts or film production, they should be able to handle a camera. Although cameras vary from model to model, the basic functionality of every camera is virtually identical. Despite the fact that his younger employees would have been exposed to digital cameras from a younger age than his older employees, and the fact that, unlike his older employees, their education would have been centered around the latest types of technology (the kind they were expected to use) he was becoming increasingly disappointed by their inability to do this rudimentary entry-level task. "They can't figure anything out, and I thought these kids were supposed to be tech savvy," he complained.

He went on to tell me about one of his recent employees who he had to fire after one day. This young man, fresh out of college with a master's degree in "media arts and design," was tasked with the very simple job of recording a three-hour speech. He emerged three hours later and confidently handed his employer his recording. It was ten minutes long.

"What happened to the other two hours and fifty minutes of the speech?" the employer inquired.

"The battery died," he answered.

Dumbfounded, the employer asked, "Why didn't you plug it in?"

"Uh, was I supposed to plug it in? You never told me that; how was I supposed to know?" the young man with a master's degree in video production responded.

"Why do you think I gave you the power cord, an extension cord, and set your camera up next to a power outlet?" His question was met with a shrug of confusion. At this point, more astonished than mad, the employer added, "At the very least, why didn't you come find me when the battery died and ask me what you should do about it?"

Without missing a beat, the young man replied, "You didn't tell me to do that."

Presumably, while working with cameras and electronics for his master's degree in the field, he would have encountered the problem of a dying battery. Considering 98 percent of his generation have cell phones, this young man undoubtedly at some point in his life had to plug in a phone or a host of other battery-operated devices. Yet he could not deduce that a similar type of electronic device, which had an almost identical power cord, would operate under the same principles as all other battery-powered devices. To him, it was the employer's fault for not telling him that his camera needed to be plugged in.

This is precisely what we as teachers are seeing in the classroom. Over the last decade our directions on every assignment have had to be made increasingly more detailed. For example, at the beginning of every class, I post four to five questions on the board under the big bold heading *Warm-Up*. Ten years ago high school students could infer that I intended for them to answer these questions, and even could come to the conclusion that they were to be done on the piece of paper I had given them with the words "Warm-Up" written across the top, with corresponding blanks for their answers. As soon as the bell rang, we'd go over them, with the entire exercise taking five minutes.

Today, I have to write, "Please answer each of the questions on the handout entitled 'Warm-Up' that I provided at the beginning of the unit." Because most students don't think to look at the board, I have to tell each one to "read the board for directions" as they enter the room. When too many students were "answering" the questions with "I don't know" (they would actually write that in the blank) I had to add to the directions, "Please look up any information you don't know in your notes." When I continued to get too many I-don't-knows I had to add the exact place in their notes where they could find the answers. These are not the digital natives we were promised.

## The Brain: Use It or Lose It

Several studies have shown brain atrophy in young people with an overdependence on technology. Two studies done on teens diagnosed with Internet addiction disorder both showed significant atrophy of intrabrain connective pathways. Another study done on online-gaming addicts also showed significant white and gray matter atrophy in several regions of the brain including portions of the frontal cortex. And yet another study done on Internet-addicted teenage boys showed they had a significant decrease in the thickness of their cerebral cortex. This would seem to indicate a decline of neurons in this region and would lead to diminished cognitive abilities.

The mystery of our disappearing brain and brain function starts to become clearer as neuroscientists begin to make sense of these three-pound lumps of gray and white matter lodged into our disproportionately large heads. With recent advances in technology (MRIs and PETs, in particular), our understanding of the workings of the brain has greatly improved over the last several decades. It's been these technologies that have led many neuroscientists to conclude that overusing screen technology can be harmful to a child's mental development. As with many of our cognitive abilities, the logical and critical thinking parts of our brain do much of their developing in the early to late teen years. Although it seems now that this process is ongoing for much of one's life, the importance of this time frame is critical.

Beginning in infancy, a child's brain begins the process of hardwiring itself. The ten billion neurons start forming nearly one hundred trillion connections to all the different motor and sensory inputs of the body through a series of connections known as synapses. These nerves and synapses act as two-way bridges that relay information (through what is known as a neurotransmitter) from one portion of the brain to another.

There are over eighty billion of these nerves connecting virtually every part of the brain to another. There is not just one pathway connecting one region to another. As the brain develops in infancy, it is uncertain what type of mental stimuli the brain will encounter or what it will be expected to do with them. So to play it safe, the brain overcompensates by allowing pathways to connect virtually everything in the brain to everything else. This allows the brain to develop in an infinite number of possibilities in which it is tailored to perfectly fit its environment. It is this interconnectedness of the different regions of the brain that gives people their intelligence.

Herein lies the problem. The brain starts off with too many of these synapses and pathways. It simply does not need them all. These pathways use energy to stay open. The brain is a very efficient machine and it hates to waste energy. So from the age of two to five, a toddler's brain starts realizing what pathways it is actually going to need, and what pathways it does not. If the brain rarely uses a particular pathway, it simply closes and discards it. This is an important developmental process known as *pruning*. Just as a gardener would cut back the dying branches of a bush to let the living ones become stronger, the brain removes underused synapses to strengthen the synapses it does use. If the brain did not do this, the excessive number of synapses would be too much for the brain to handle. Pruning allows the brain to focus on the pathways it actually needs. This creates the use-it-or-lose-it principle of cognitive ability. As the brain prunes away rarely used synaptic connections, it loses the potential to realize that particular cognitive ability later in life. A child who rarely uses imagination in childhood will be unimaginative later in life.

This is how childhood experiences shape cognitive abilities. The importance of pruning in toddlers has been known for some time. However, more recent discoveries have shown that teens age twelve to eighteen also go through a significant period of growth and pruning in the prefrontal cortex. Despite the fact that the brain is roughly 95 percent of its adult size by the age of six, the prefrontal cortex goes through an increase in density of gray matter as it continues to hardwire itself during a later growth spurt.

This late-stage pruning makes the teenage years vitally important to the development of higher-level cognitive skills. Dr. Jay Giedd, a neuroscientist at the National Institute of Mental Health in Bethesda, says, "If a teen is doing music or sports or academics, those are the cells and connections that will be hardwired. If they're lying on the couch or playing video games or watching MTV, those are the cells and connections that are going to survive."

It also means that if a teen underutilizes the portions of the prefrontal cortex used to solve problems or think critically, the brain will view this portion as unnecessary and begin the process of removing the synapses connecting it to other regions of the brain it does use, such as the portions used for gaming and social media.

How is it possible the human brain could view the problem-solving regions as unnecessary? Rarely today do teens come across a question or a problem that cannot be instantly answered or solved by an app or electronic device. If a student is facing a challenging critical-thinking question, all she needs to do is type it in to a search engine, and she will most likely come across a variety of sources to answer the question for her. Contrary to the digital native claims, millennials aren't using these answers as a starting point to explore solutions to new problems. They're using them to answer homework problems as quickly as possible and go back to Netflix, their game, social media, or porn.

## Dumbed-Down Curricula, Dumbed-Down Students

Because modern technology is so capable of solving our problems for us, many protech educators are advocating for replacing older curricula focused on traditional thinking skills with more modern "twenty-first century skills"—in other words, lessons on how to use screen-based technology. Many school systems today have or will purchase a laptop or other digital device for every student from kindergarten to twelfth grade. The main idea is that teachers can then give lessons on how to be productive digital citizens.

First of all, this shift in educational practice completely contradicts all the claims about digital natives. If they're already so well versed in technology, why spend school time teaching it to them? They're the ones who are supposed to be teaching digital immigrants about technology, not the other way around.

Second, what lessons in technology are we going to teach five-year-olds that will still be relevant when they enter the workforce sixteen years later? This would be the equivalent of elementary schools of the '90s replacing time spent teaching reading, writing, and arithmetic with lessons on how to program VCRs. It's laughable to think public schools can keep up with the lessons of modern technology. Even lessons taught in high school are likely to be obsolete by the time their students enter the workforce. My senior year in high school, I took a class on C++ computer programming. On the first day of class, the very first thing our teacher told us was "This is the first and last year we will be offering this course. Everyone uses JAVA now." This is because curricula (if they're any good) take years to develop. Textbooks are typically used for a decade before they're replaced.

Schools cannot afford to keep up with the ever-changing world of technology.

Finally, modern technology is so intuitive and user friendly that it doesn't need to be taught. Virtually anyone with any common sense can figure it out. The following example came from a teacher I talked to from North Carolina who works with severely mentally handicapped students. One day, when collecting iPads she had used for a lesson, she noticed one of her students, a seventeen-year-old boy with a third-grade reading level and an IQ of forty-nine, had managed to bypass the school's parental blocks and had downloaded a video he had searched for entitled "How to Perform Coitus." You can't make this stuff up.

Children are spending less time doing mentally rigorous schoolwork, and exponentially more time engaged in mindless leisure activities. Their brains, therefore, become more geared for leisure activity than deep thought. In 2010 teens were averaging sixteen minutes a day to help them complete schoolwork. This is barely enough time to type a question into Google and copy and paste the answer into their homework assignment.

Unfortunately, copying someone else's answer has never been an effective way to learn. At least predigital forms of copying required more interaction with the content, with an increased likelihood of reading it. Today, with a few keys strokes, entire paragraphs can be lifted without ever even reading the content. Andrew Keen, an outspoken critic of the digital native claims, characterizes the younger generation in his 2007 book *The Cult of the Amateur* as "intellectual kleptomaniacs, who think their ability to cut and paste a well-phrased thought or opinion makes it their own."

From a teacher's perspective, these observations are spot-on. Luckily for teachers, as the level of effort required to cheat has decreased, the ease of spotting a cheater has increased. Students forget (or don't know how) to remove the hyperlinks embedded within a Wikipedia article they copy, or they don't know how to get rid of the light-gray background that transferred over with it. Papers are often littered with font changes, highlighting everywhere the student copied from a different source. I have had students accidentally hit the paste function twice, leaving two identical plagiarized paragraphs in a row.

Even without all these red flags of cheating, which further illustrate just how non–tech savvy digital natives are, it still is painfully obvious when a student isn't doing his own work. Students can't hide a tone change. Plagiarized statements leap off the page as students jump in and out of informal and formal tones. Recently, I had given my students a critical thinking question to take home and answer with a formal response. I try to make these types of questions require deeper levels of thinking by leaving them open-ended and including opinion, rather than simple fact recall. When reading one student's response, I was struck by how his introduction sentence was riddled with typos and informalities, like his opening line, which read, "This is my paper." He then went on to answer the question in a brilliant, formal, and educated manner. What I found particularly puzzling was how his response answered my obscure question perfectly. *There's no way he found this answer on the Internet*, I thought to myself. He would have had to spend hours going through different websites just to find such a perfect response.

So I typed a few of the key words into a search engine, and the first thing to pop up was the verbatim text of his response. Apparently, unable to quickly find the answer to the question, this student had typed it into Reddit—a website that's essentially a collection of specific themed discussion boards. Minutes after he had posted the question, several people had gone to the trouble of answering it for him. The answer he chose to copy and paste into his assignment was written by someone claiming to be a history professor—which, judging by the caliber of the response, he probably was.

Ed-tech advocates would look at this incident as one of the biggest upsides to modern technology. Here, a fourteen-year-old high school student from an East Coast suburb was able to "discuss" ancient Mesopotamia with a professor of ancient Mesopotamian culture in California. Technology has given this student access to expertise in a specific topic that goes far beyond anything he can get from me or his textbook.

Let's assume for a second that the person who answered his question was who he said he was. (I have enough knowledge on the topic of the ancient Mesopotamia to appreciate the brilliance of this particular response. A student, however, who is just beginning to learn about the topic typically doesn't have enough background knowledge to discern between a brilliant response and a totally nonsensical one generated by some weirdo in his parent's basement trolling the Internet for his own amusement.) The ed-tech advocate misunderstands the nature of this interaction. My student didn't engage this professor in a back-and-forth discussion of the topic. He didn't "collaborate" with the professor on a proper response. He didn't sit in on one of his lectures or ask the professor any insightful follow-up questions, or even provide a perfunctory "thnx." My student simply copied and pasted the question from his assignment into Reddit, checked back a few hours later for a response, and copied and pasted the most popular response back into his assignment.

I'm not even sure my student fully read or understood his answer before he turned it in. When I asked him to read it aloud, not only could he not pronounce or define a third of the words, he couldn't even give me a general sense of what he was saying. My student had gained nothing from this exercise other than the "right answer." This is because copying someone else's ideas is not a useful method of learning or thinking. It doesn't matter if it's a student copying test answers from the kid sitting next to him, copying a paragraph from the encyclopedia, or copying a college professor's response to a specific question.

This is why we as teachers take "academic integrity" so seriously. Besides being unethical, copying someone else's work requires little if any thought. But deep contemplative thought is required for learning. If all it took to understand a concept was a quick Google search, highlighting, copying, and pasting, schools wouldn't be necessary. Kids could spend a few hours copying and pasting Wikipedia pages for a few months and have all the knowledge they would ever need.

What I found most disconcerting about this incident was the student and his parents' response to it. Within minutes of discussing this with the student, his parents reached out to me in disbelief that I was expecting their son to do the assignment again. Like their son, they failed to see how this attempt at the assignment was unacceptable. From their perspective, he had done the assignment. If anything, they were proud of their son's ingenuity in using the Internet to find the correct answer. For the son, this method of copying and pasting was all he had ever known. Being a digital native, this is just how critical thinking assignments like this were done. To him, not only did he fail to see how using someone else's response was unethical, it was unfathomable to him that I expected him to generate the answer on his own. "How was I supposed to do it?" he finally asked me. It was as though it had never occurred to him that he could think about the question on his own.

What the parents failed to understand is that their son's attempt at completing this assignment did nothing more than reinforce his skills on how to use the website Reddit. But I wasn't trying to teach him how to use Reddit, a very simple website that requires no thought or skill to use. I was trying to get him to use his brain and think about a specific concept in world history. My version of the assignment required him to think deeply about the things he knew about Mesopotamia in order to generate a response. This process of thinking is what cements the learning in his mind. It would have been more productive for him to think really hard for a few minutes and generate his own poorly worded response than to think very little and generate a college professor–quality response. Rather than using modern technology as a starting point to increase human intellect and spawn new and creative responses, in this instance it was used as a means to an end to replace intellect.

With knowledge so devalued, students in this generation could potentially spend their formative years learning how to be wonderful copy-and-pasters. This is where Madison, a ninth-grade honors student, began to struggle. She was a heavy tech user and firmly believed that she shouldn't have to learn anything she could look up. She had always been a good student in elementary and middle school, but her grades began to slip early in her ninth-grade year. The first quarter, she was getting mostly Bs. Her parents were concerned, but they attributed her decline to the typical teenage adjustments to high school. But by third quarter, her Bs had become Ds, and she was starting to fall apart. To her, school seemed like an unnecessary burden.

After much insistence, she eventually came in for additional help. The first thing I asked her was "Why do you think you're struggling?"

Madison gave me a familiar response. "I'm just not a great test taker," she said with a dismissive shrug.

This is a cop-out. I hear this all the time. Virtually every student who consistently struggles blames it on not being a good test taker. There is no doubt that some students have genuine test anxiety and freeze up on test day, even though they know the material. But that is pretty rare. In reality many students have heard adults throw the phrase around, so they think it has some validity. And more important, it absolves them of any responsibility for their own performance. They fail because they're not great test takers, not because they didn't learn enough of the material on the test to be successful. The vast majority of students who make this claim struggle because they are underprepared, not because they're genetically predisposed to fail tests.

I pointed out to Madison that she had been an A student up until that year—so she must have done well on tests before high school. "So what's changed?" I asked her.

"It's just that high school teachers ask a lot of trick questions that are really unfair," she said in a reluctant, cautious tone. She was afraid of offending me, but my skin has grown extremely thick over the years. She needn't have worried. I have heard this complaint with increasing frequency over the last few years. Although the structure of my questions had changed little over the last decade, more students were complaining that they were "trick questions." They weren't. I pride myself on writing fair questions that cover major concepts. I don't ask questions on minutiae and I don't ever try to trick students into choosing the wrong answer. No reasonable teacher makes a regular practice of that. But I knew what Madison was saying.

In more traditional education, the level of rigor increases as a student moves up in grade. In the lower grades, many of the questions are simple fact-based questions, requiring the lowest level of thinking. In teacherspeak, we call this level of thinking "knowledge" or "comprehension." In most cases, answering a question like this requires a simple recitation of a memorized fact. It is these lower levels of thinking that modern technology is particularly good at. Google "Who was the third president of the United States?" and the correct answer will appear within a fraction of a second. Madison's acceptance of this type of learning had solidified by the time she made it into the ninth grade. Why should she have to memorize which Babylonian leader codified the first set of laws when she could find this out with a quick Google search?

This mentality shaped how she viewed school. Rarely did she engage in class and she wouldn't do any of the classwork or homework. To her, this was busy work and not a valuable use of her time. Her strategy was to start preparing the night before the test. She never took notes in class, figuring she would just download the PowerPoint presentations later. She'd take the list of terms I would give as a study guide, make digital flashcards on Quizlet, and try to memorize as many as she could the night before the test. After all, this is what had worked for her in the past.

And this is why she found high school test questions so tricky. No longer was she being asked questions such as "Which Babylonian leader codified their first set of laws?" Now she was being asked to use higher levels of critical thinking with questions such as "How is the Babylonian code of laws similar to the Ten Commandments of the Hebrews?" This isn't a "trick question," but her memorized fact of "Hammurabi wrote the Code of Hammurabi" did little to answer it.

Madison's problem was that she didn't have the requisite knowledge necessary for the higher levels of critical thinking. Her strategy of "I'll look it up when I need it" had left her with an extremely shallow grasp of the material. She only knew random facts sprinkled throughout an entire unit. This is because she began to think like a poor man's Google. When trying to come up with answers, she could only generate quasi-related facts. Without any context, she struggled to connect these facts in any kind of meaningful way. In order to think critically about something, one must first understand the basics.

This problem created by the disjointed "Google thinking" manifested itself most clearly in our vocabulary assignments. For each unit, I give my students a list of roughly ten to fourteen important vocabulary terms that pertain to the topics we're discussing. I provide the students with the definitions, and then have them complete a series of activities in which they have to use the words in a variety of contexts. Many of these words are conceptually related to one another. For example, in our unit on the Neolithic Era, we will have the terms *agriculture* (the process of domesticating plants and animals), *civilization* (the advanced stage of human society built from permanent settlements), and *Neolithic Revolution* (the era in which many groups of people transitioned from hunter-gatherer societies to agrarian ones). Madison (along with most of her peers) excelled at committing these definitions to memory. I could ask her what each word meant and she could give me a definition pretty similar to the one I provided for her. However, she struggled to connect the dots. She couldn't conceptualize how the discovery of agriculture was known as the Neolithic Revolution, which allowed for the development of permanent settlements, which led to civilization. There was no fluidity in her thinking. The thoughts she did have reflected the staccato-like fashion in which she used Google. "What is agriculture?" "What is the Neolithic Revolution?" "Define civilization."

This method of thinking is very much reflective of how digital natives interact with technology. They answer questions one at a time while jumping from concept to concept. They are likely taking breaks between concepts while they "multitask," further disconnecting the related concepts. The older methods of obtaining information, such as lectures and paper books, follow a linear, sequential pattern. This is a more ideal way to learn. Learning sequentially not only can give students the knowledge of the individual concepts, but it also allows them to connect that knowledge to the broader picture.

Although Madison had knowledge of the definitions, she had no understanding of them. She wasn't getting what they meant or why they were important. Had she participated in the lessons in class, these connections would have been abundantly clear. But remember, Madison was spending her time in class engaged with her technology, waiting until the night before quizzes and tests to look up the information she was missing. Unable to make sense of this information in how it applies to the bigger picture made these terms seem meaningless. The brain has a habit of discarding seemingly meaningless, unimportant information. Technology doesn't encourage one to jump into critically thinking about things of which one has no understanding. Because of this, not only was she missing the bigger concepts, but the facts she was trying to memorize weren't sticking because they were totally void of any meaning.

Regardless of how easily accessible information is, there will always be a need for basic internalized knowledge and understanding. This base of knowledge is what allows us to make sense of the world and helps us solve the problems we encounter in it. The greater the extent of this knowledge, the greater our ability to solve problems. Students struggle to research on the Internet not because they can't figure out how to type words into Google but because they lack the fundamental knowledge to get started. In order for a student to find a primary source on the Crusades, he would first need to be able to define "primary source" and "secondary source." It would also help if he knew what the Crusades were, and the names of specific key figures, places, and events. From this basic knowledge, he would then have to employ higher levels of thinking. He would have to compare his understanding of the definition of a primary source and secondary source against an actual source to see which definition best applies. He would have to understand the role the key figures played in the Crusades to evaluate their sources for bias.

This is why parents and educators need to ignore the digital native myth and forget the claim that you can just throw a laptop or iPad at them and have them solve problems on their own. There is still a need for an incremental approach to teaching and learning. Rather than spending their days teaching students how to use different devices and programs that will be obsolete by the time they graduate, schools need to maintain their traditional goal of providing students with a foundation of knowledge.

If you consider the fact that technology is designed to solve problems for us, it should only be introduced after students have mastered the ability to solve problems on their own. Otherwise, technology works to mask their lack of understanding. For example, if you were to give a student a calculator moments after teaching her multiplication and division, chances are she would stop working through the problems on her own and start depending on the calculator to do the work for her. The calculator would become more of a crutch because her own multiplication and division abilities would never fully be formed. This gap in her understanding would become more apparent later when teachers tried to build on these simpler concepts with more advanced ones, such as algebra or geometry. And eventually, the student would reach a point, much like Madison did, where she struggles with new, more challenging content and concepts.

If, however, you allow that same student to work through the simpler math concepts without any technological aid, she will eventually master the skill on her own. This may be a long and painful process. And it may feel unnecessary—with high powered calculators on phones, who needs to be able to do this? But having this understanding will make future learning possible. She will be able to conceptualize new ideas better and make sense of them because they fit into her existing mental framework. If at that point, when teaching these more advanced concepts, you introduce a calculator, it's no longer a crutch. It's helping her to perform a task that she understands, just in a faster, more reliable way.

The solution for Madison was a "reboot." She had to go back to the basics and start working to master the fundamental concepts before she could start thinking critically about them. She started paying attention in class so that she could better answer her nightly critical-thinking questions for homework. In history, her grades eventually turned around. In math and science, it took longer. Her parents ended up getting her a personal tutor to reteach skills she should have mastered earlier. But once she did master them, higher-level thinking questions gave her little trouble, and her tests no longer seemed so tricky.

## The Sounds of Silence

The final suggestion I made to Madison was seemingly insignificant. Like many of her peers, she always had her earbuds in. Before school, after school, during lunch, in between classes, and when she'd leave class for a minute to get a drink of water. She'd even wear them in class if I let her. Her mom told me she pretty much had them in from the time she woke up until the time she went to bed. She loved listening to music, so much so that she hardly spent a minute a day in silence.

As it turns out, this is a problem for the developing young mind, particularly for their critical and creative thinking. A study by Dr. Sandi Mann and Rebekah Cadman from the University of Central Lancashire found that people do their best thinking when their mind is inactive. Mann and Cadman found that when doing boring, less engaging activities, participants actually produced creative ideas that were both higher in quality and quantity. In the study, people were asked to think about different uses for a plastic cup while performing one of three tasks with increasing levels of monotony. The people performing the most monotonous task (reading a phone book) generated the greatest number, and the most outside-the-box responses.

Another study conducted by Karen Gasper and Brianna Middlewood of Penn State University found similar results. In their study they showed different films designed to elicit a specific emotion, such as boredom, elation, relaxation, or distress. After watching one of the films, the participants were then asked to generate responses that fit within a particular category, such as "types of vehicles." The people made to feel bored were more likely to generate creative responses, such as "camel" or "Segway," whereas the more stimulated participants tended to pick the more tried-and-true responses such as "car." When asked about her findings, Dr. Gasper said, "It results in you trying to approach something that, in this case, is more meaningful or interesting. It encourages people to explore because it signals that your current situation is lacking so it's kind of a push to seek out something new."

Digital natives don't allow themselves to be bored. They fill any downtime with music, videos, games, or social media. A substantial majority of students today sleep with their phones either next to them or under their pillows. They spend the moments before they fall asleep and the moments after they wake up on their phones. Car rides, walks, or any time spent alone is spent engaged in some type of digital activity. Bus stops are filled with children staring at their screens. One student told me he puts a Ziploc bag over his phone so he can watch TV shows while in the shower. These are all missed opportunities for quiet reflection. Downtime is typically when people do their best thinking and work to solve their daily problems. The digital native's inner monologue has become a stranger to them, and a major reason why critical thinking is on the decline.

## TAKEAWAYS

- Encourage children, especially young ones, to play in ways that cultivate imagination. Imaginative play is one of the best cognitive exercises for a child. Playing house, school, or other imaginative role-playing scenarios are great ways to foster the innate creativity all children have. Reserving childhood for play will have huge educational benefits down the road.

- Teach kids how to solve problems on their own before introducing technological aids. Only once they've mastered these skills should technology be introduced as a means of augmenting their existing abilities. Introducing technology before they've mastered these skills on their own only works to create dependency on technology, and it makes future learning more difficult. For example, don't allow your child to use a calculator until she has mastered multiplication and division without its aid.

- Instill in kids the importance of knowledge. Challenge kids to think about what they're being asked to do at school and on homework before they dismiss it as "busy work" that isn't worth their time. Help them see that they must *know* things before they can *do* anything. Tell your kids about your own struggles to learn—in school, work, or wherever—and point out personal examples of how you weren't able to master anything until you had mastered basic knowledge.

- Allow kids to be bored. Make cars technology-free zones. Take their phones out of their rooms. Allow them to sit in silence. Bored kids—whether two or seventeen years old—can be incredibly creative. That creativity is critical if students are going to develop problem-solving skills. When they come to you with the all-too-familiar refrain "I'm bored!" give them something to think about. I am not naive enough to think that if you respond to a bored kid, "Well, just try thinking!" they'll reply, "Great idea!" and happily skip away. However, we all have long-range projects and situations in our lives and homes that need some thought. Enlist their help in figuring out creative solutions. One of two things will happen: either they will dig into the project and help, or they will say, "Wow, that's the worst idea I've ever heard," and they'll figure out a new way to not be bored. Either way, it's a win.

These suggestions can only take us so far. Students can only learn how to think and solve problems creatively if they can block out the noise of the world and truly focus their mental energy. This is the focus of the next chapter.

4

# Learning to Focus
# in the High-Tech World
# of Distraction

*They're pretty much mental wrecks.*
—STANFORD SOCIOLOGIST CLIFFORD NASS,
DESCRIBING PEOPLE WHO FREQUENTLY MULTITASK

A GLANCE AT THE CLOCK confirmed my fear. It was getting dangerously close to 6:00 PM. I was in a frenzied panic as I dashed back and forth through my classroom trying desperately to finish preparing for the evening's events. Typically, you would never catch me in my classroom around 6:00 PM. But this was a special circumstance. It was back-to-school night.

I glanced out the door and saw the first wave of parents nearing. These are the parents to be feared. Side by side, they casually stroll the halls while pointing to the artwork and posters adorning the walls. But this is all an act. They are only pretending to be aimlessly wandering as they "kill time." "Oh, it's right here, honey," one parent will say to the other as they "accidentally" stumble into my classroom. "I thought we'd never find it!"

This was no accident. These parents were on a mission. They understood that this was the perfect opportunity to corner poor unsuspecting teachers into an extended conversation about their child. This conversation normally goes something like this: "Hi, we're _insert name_'s parents. How is _insert name_ doing? [He or she] just loves your class. It's [his or her] favorite subject."

My mind races as I frantically try to remember who _insert name_ is. Two weeks into school and I haven't yet put hundreds of names to hundreds of faces. It helps if there is a strong family resemblance, which there usually is not. And even if I could attach the name to a face, what possible insight could I give after only a handful of classes? Should I compliment their child's good attendance, clean clothes, or purchased school supplies?

I froze as I saw a face pop into my window and the doorknob begin to turn. I had been discovered.

"Hello! Are we early?" a well-dressed couple inquires while grinning from ear to ear.

_Of course you are,_ I think to myself. But what I say is, "Oh no, not at all! Please come in and have a seat!"

I hope they will just take a seat and begin conversing between themselves. But it's not my night. Rather than sit, they make a beeline for me.

"Hi, we're Jayden's parents, and he just always talks about how much he has already learned this year," they say as they move in for a handshake.

This was clearly a lie. So far we've only learned about map projections, and who would possibly go home and be excited to talk about map projections? Nobody.

_Wait a minute,_ I think to myself, _I'm in luck. I know who Jayden is._

Like any student whose name a teacher knows this early in the year, the only reason I was familiar with Jayden was that he had exhibited some bizarre behavior. He would walk down the hallway wearing oversized headphones (the ones designed to block out all outside noise) while reading from his e-reader. Now, navigating our crowded hallways can be a challenge for the most focused person, and it is always amusing to watch students try to do it while looking at their phones, a laptop, and in this rare instance, an e-reader. But Jayden was the worst at it. He was oblivious to the world around him and would regularly walk into people, doors, and trash cans. Students would try to get his attention or engage him with little success as he fumbled his way to his seat.

"Oh, Jayden? Of course!" I say, "He's a wonderful student."

I'm being nice. He is not a wonderful student. He has been glued to his electronic devices every class until I knock on the desk to get his attention. And even after I finally get him to focus, he can only do so for a few moments before going back to his e-reader, phone, or whatever gadget I have not yet explicitly told him to put away.

The parents' lingering smiles and unblinking eyes made me realize they were waiting for a follow-up compliment. I'm lucky to come up with, "He is great at self-advocating!"

This wasn't a lie. Jayden had already e-mailed me twice to ask me what that particular night's homework assignment was. Although I typically would applaud students taking the initiative to reach out and ask questions, I find the increasingly common "Hey, what's r homework?" e-mail particularly annoying. This is because I hand each student a calendar outlining all assignments and their due dates at the beginning of every unit, post a copy of said calendar at the front of the classroom for students to have access to in case they lose theirs, post an electronic version of the calendar on the class website alongside their actual homework assignment so that they can download it anywhere and anytime, and then remind them about what is due at the beginning and end of every class as I point to where I've written the assignment on the portion of the board permanently labeled "tonight's homework assignment." But like so many of his peers, he openly laments, "It would be a lot easier if you just e-mailed us our homework every night."

Now, I don't know what came over me this particular night. Back-to-school night (sometimes called curriculum night) is not the time to address real issues with students. But in a moment of bravery and because they were still standing there wanting more information, I decided to steer the conversation into a noncomplimentary direction.

"If we could just get him to take his headphones off for the first few moments of class and put down the e-reader," I offered, "I think he'd take in some of the things going on in class, like what assignments are due and when, or some of the content we are going over."

Unflinchingly, the parents turned my statement into an affirmation that their son was in fact, as they suspected, a great student. Beaming, they add, "Oh, he's been doing that for years. He loves to read one book on his e-reader while listening to another book-on-tape. He's doing that while listening to everything you say, too."

I couldn't believe it. Did they just say Jayden is reading two books at the same time, while listening to me? Every time I would call on him, he'd reply with "What question are we on?" or "Where are we?" The class would laugh when he would ask a question I had literally just answered moments before. "He literally just said that," one of his annoyed classmates would chime in. These parents clearly had no idea just how tuned out their son was at school; they believed instead that he was a brilliant multitasker.

## Multitasking Myths

This is a common myth about the digiLearner. At this point, some-one has probably tried to convince you that your children's brains work differently from your own more primitive "digital immi-grant" brain. Through their constant exposure to multiple forms of media, digiLearners have actually trained their brains to split their focus and simultaneously complete multiple tasks—"multi-tasking." They can do their homework while watching TV, while on Facebook, while texting their friend, while listening to music. Jayden's parents ended our conversation with "It's just how kids today focus."

After years in the classroom, demanding kids focus and pay attention has led me to one simple conclusion—multitasking and focus never go together. As Dr. Edward Hallowell, a psychiatrist who specializes in attention disorders and author of the book *CrazyBusy*, defines it, multitasking is the "mythical activity in which people believe they can perform two or more tasks simul-taneously." I'm not talking about simultaneously completing any two activities. I'm focused less on instinctive activities like breath-ing, walking, or chewing gum and more on complex, effortful activities such as listening to a lecture while playing a game or reading a book.

"Multitasking" has become an important part of the digital native's life. According to a Common Sense census, half of teens are on social media while doing homework, more than half are watching TV, 60 percent are texting, and 76 percent are listening to music. Because digiLearners are spending 30 percent of their digital time exposing themselves to multiple forms of media, or multitasking, the average teen can consume almost eleven hours of media content per day. And it's not just the "average" student wel-coming these digital distractions. A survey of the top 25 percent of

Stanford students found that they are using four or more different types of media at a time.

The digital native sees no problem with this. Nearly two-thirds of teens surveyed thought watching TV while doing homework had no impact on the quality of their work. Similarly, over half didn't think being on social media affected levels of concentration, and half thought that listening to music actually improved the quality of their work.

But their assessment of their own abilities is far from accurate. According to David E. Meyer from the Brain, Cognition, and Action Laboratory of the University of Michigan, "Multitasking is going to slow you down, increasing the chances of mistakes." He went on to say that his research showed that "disruptions and interruptions are a bad deal from the standpoint of our ability to process information." And "processing information" is exactly what we as teachers are expecting our young people to do when they're working on their homework or paying attention in class. Teachers don't assign worksheets filled with math problems because they have a quota of solved equations they need to meet. They want their students to take in the information and work through the problems in a focused and thoughtful manner to better internalize the skills and concepts so that they may "learn" them and recall them later.

Activities such as listening to a classroom discussion or doing homework can be done passively (requiring little focus) or actively (requiring lots of focus). When passively listening, a student can get a general sense of the topic being discussed, and might even be able to repeat the last sentence the teacher just said. When passively doing an assignment, he can generate answers that may be somewhat satisfactory. But what he can't do is fully process and internalize the information being conveyed. When the mission of teachers is to convey important information in a deep and meaningful way so that the student may learn it for a lifetime, teachers require the student to be actively engaged in the lesson.

A study done by David Strayer, director of the Applied Cognition Lab at the University of Utah, discovered that 98 percent of the population cannot effectively divide attention between two tasks at the same time. No research has shown that anyone can do three or more tasks at a time.

But you don't have to be a professor at a cognition lab to know this to be true. Think about this. If digiLearners do in fact have the ability to train their brains to have a superior ability to multitask, you would expect teens today to have the lowest rate of distracted-driving accidents. Applying the logic that today's teens have been multitasking since they were toddlers and are therefore better at it, one could assume that teens can safely navigate the roads while simultaneously texting, e-mailing, or looking for another song on their iPod (or all four). However, statistics released by the US Department of Transportation on incidents of distracted driving show the exact opposite to be true. The youngest drivers studied (ages eighteen to twenty) had the highest rates of accidents related to distracted driving. The number of these incidents almost doubled that of the next-highest age demographic (ages twenty-one to twenty-four) and more than quadrupled that of those over sixty-five. The rate at which these accidents were caused by phone use was the highest for the youngest demographic, and 13 percent of all accidents were caused by attempts of the young driver to multitask. An example of this is our school parking lot ten minutes after the final bell rings. Teachers and parents take their lives into their own hands when they enter the parking lot full of teen drivers with distractions aplenty. Digital natives can navigate their way through a set of math problems while texting about as well as they can drive while texting. That's to say, they can't.

## Pay Attention to This

To understand why almost everyone lacks the ability to truly multitask, we must first understand the human brain's ability to focus. Many neuroscientists today believe that our brains have a finite and limited amount of focus. People lack the mental facilities to give their full attention to everything going on around them. Two Harvard neuropsychologists, Christopher Chabris and Daniel Simons, best illustrated this in their now-famous study. They showed their subjects a video featuring two groups: one group of three people wearing white T-shirts and another group of three people wearing black T-shirts. Each group had its own basketball, which they would pass back and forth between the members. Chabris and Simons gave their subjects the effortful task of counting the number of passes made by the group wearing the white T-shirts. Because their attention was so focused on the basketball of the white T-shirt group, most people inadvertently tuned out all irrelevant information, which in this case was anyone wearing black. Fifty percent of the participants did this so effectively that they failed to notice a person who appeared on the screen dressed in a black gorilla suit. To make the gorilla more obvious, Chabris and Simons had him jump up and down while waving his arms. This display was so obvious that 0 percent of people who were not given the task of counting basketball passes missed it.

Neuroscientists call this "inattentional blindness." They liken attention to a spotlight illuminating a stage. When narrowly focused, a more intense light highlights all the minor details in a small area. While it does this, however, it leaves the rest of the stage in the dark. When one's attention is focused like this, it allows him to count the number of times a basketball changes hands, make sense of reading an article on the structure of DNA, or understand a lesson on how to convert fractions into percentages. But it also forces him to miss everything else going on outside of his narrow focus. Our attention can also have a broader focus, spreading a less intense light over the entire stage. Someone with this broader-focused attention couldn't count the number of times a basketball changes hands but would notice a gorilla appearing on the screen. Likewise, in a classroom, a student may notice that the teacher is talking about DNA and notice Tim as he came in late to class. However, simply recognizing that the lesson was about DNA is not enough if the student can't recall any of the specific details from it.

From a teacher's standpoint, we see the effects of inattentional blindness almost every day. Nothing illustrates it more than giving directions for a test. When I give a test, I start by passing out the answer sheets (Scantrons) that students use to bubble in their answers. The Scantrons we use have a place for their name, date, and the subject of the test. Once every student has one, I make the announcement that, as always, I've written the test subject and the day's date on the board for them to copy. Pointing to the giant block numbers and letters, I make the same joke, "All you need to know is your name, which if you don't, you have bigger problems than this test." No one laughs.

What follows is easily the most frustrating part of being a teacher. As soon as the answer sheets land on students' desks, they instinctively start filling them out. While their attention is narrowly focused on the simple task of writing their names, something they've been doing since kindergarten, they inadvertently tune out my announcement on how to fill out the lines other than the "name." "What's the subject?" will be uttered by a student no more than five seconds after I've given them this very specific instruction. At this point, I still have loads of patience because the class has only just begun. So I'll politely say, "As I just said, the subject is ——, and [because I've also written it in giant letters on the blackboard directly in front of them] it's also written right here where I'm pointing." Ten seconds later: "Excuse me, what is the subject of this test?" Now I don't care who you are, this is annoying. At this point, I've already written it once and said it twice. Doing my best to hide my annoyance, I'll say, in a curter tone, "Everyone, please listen! Please put down your pencils and look at me! I will not say this again. The subject of the test is ——, and as always, it's written right here!" And like clockwork, one student will lean to another and ask, "What did he just say?"

Students cannot do something as routine as write their names and listen to simple instructions at the same time. Yet parents are led to believe their children are capable of being on social media while listening to music while doing their algebra homework.

Chabris and Simons's study and others like it have led many modern neuroscientists to believe that besides being finite, attention is also indivisible. It is simply impossible for most people to divide their attention at any given time. This paradigm shift is evident from the reclassification of attention deficit disorder (ADD). This title is no longer in vogue because it confuses the very nature of the disorder. As we now know, people diagnosed with this increasingly common learning disability don't have less attention than the average person. What they do have, however, is the tendency to disperse their limited amount of attention over far too many stimuli at a given time. In an attempt to focus on many stimuli, they are unable to truly focus on any. This is why today the more popular name for this disorder is attention deficit hyperactivity disorder (ADHD). Their attention is "hyperactive" in that it bounces from stimulus to stimulus. As Nobel Prize–winning neuropsychologist Daniel Kahneman puts it, "You dispose of a limited budget of attention that you can allocate to activities, and if you try to go beyond your budget, you will fail."

Dr. Kahneman further makes sense of the brain's ability to focus through the mental model of "system one" versus "system two." In his model, the brain's functions are divided into two separate categories. The first of these categories is known as "system one," which "operates automatically and quickly, with little or no effort and no sense of voluntary control." The types of mental activities people allocate to their system one would be things like: adding one and one, detecting the source of a loud noise from across the room, or spelling one's name. These activities happen automatically with little focus required. Most adults don't consciously add one to another one or sound out their name. The answer appears in their mind as soon as the question is asked.

The second category is known as "system two." This system accounts for the more effortful, focused forms of thinking and kicks in when the lazy system one feels overwhelmed. The types of activities associated with system two are the more involved mental tasks. These may include things like mentally multiplying twenty-four by thirty-eight, listening to a conversation in a crowded restaurant, or comparing two historical leaders. These types of activities demand more attention. Any kind of interruption to this more intense focus can cause irreparable harm to these particular trains of thought.

The more comfortable people are with an activity, the more system one can accomplish it without the aid of system two. For example, someone learning to play the guitar is going to use her system two in order to master the left and right hand movements while trying to learn a new song. Over time, as she becomes an expert at the guitar, she can play that same melody relying entirely on her system one. Her fingers seem to think for themselves as she effortlessly moves them with no real thought.

Have you ever parked your car after your drive home from work with complete amnesia about the specifics of how you arrived there? You vaguely remember being behind the wheel and driving, but you cannot recall any particular event. You somehow managed to operate your car through a variety of conditions entirely off your system one. This is because the drive is routine for you. A new driver would arrive home with complete recollection of most turns and lane changes. But you've done it every day for an extended period and know the turns in and out and the intricacies of when to merge or what off-ramp to take. You don't need the effort of your system two, so your brain depends mostly on your system one. System one is more of a backup mode, getting you through the mundane and monotonous without having to delve into deep analytical thought.

This is multitasking's problem with regard to education. This extremely sophisticated process of focus isn't designed for the purpose of allowing students to learn while playing games or being on social media. It comes from an evolutionary need to detect potential dangers in the wild. Think of your system one as the security cameras scanning back and forth in a 7-Eleven. If no dangers are detected, the tapes get wiped clean as the cameras continue to scan. If a student's system two is focused on a game being played on his cell phone, he is only "listening" to the teacher with his system one, which is really only processing the words being uttered for potential threats that would need to alert his true focus, his system two. If no alert happens, then the nonthreatening information he heard will be disregarded as unimportant. In the classroom this "unimportant information" is the information the teacher is trying to convey to the students. The focus of the effortful thought inherent in system two is required for true learning. The students who carelessly rely on their system one throughout an entire class are likely to remember as much of the lesson as you do about your drive home.

People can set their system one for a variety of situations. For example, rather than scanning the horizon for a lion, they can set their system one to alert them when a shock of white hair walks by as they are looking for their grandma at the airport. Similarly, after making an important point during a lesson, I typically say, "That will be on the test." Like a pack of prairie dogs, the students' heads collectively pop up as they shift their attention from their cell phones to me. Their eyes will dart back and forth as they wait for me to reiterate the point. Once they realized they've missed it, they will turn their heads to their neighbor and whisper, "What's going to be on the test?"

In class, students set their system one to scan the lesson to alert them to important information. So far degraded is their attention span, however, that these alerts are pretty much limited to the teacher directly stating, "This is important." But saying "This is important" before every statement isn't a practical method of teaching. I've had many students admit that they spend entire blocks of "boring" classes on their phones and that they are completely unable to recall one thing that went on. Even they will admit that they learned nothing during that block of time.

A University of California, Los Angeles, study illustrates why this may be. When using MRIs on subjects to determine the impact of multitasking on their ability to learn, researcher Russell Poldrack discovered that multitaskers actually use a different region of their brain when compared to the focused subjects. While the focused learner uses the hippocampus—the region of the brain known for creating deep, lasting learning—the multitasking learner uses regions more closely associated with the automatic system one. According to Poldrack, "Even if you learn while multitasking, that learning is less flexible and more specialized, so you cannot retrieve the information as easily." This, in part, helps explain students' trouble thinking creatively and applying their learning to a variety of scenarios outlined in chapter 3.

DigiLearners mistake this ability to do one thing and tune in when needed as multitasking. However, rather than multitasking, they're doing what neuroscientists call "multiswitching" or "task switching." Rather than focusing on both stimuli simultaneously they are shifting their full attention from one stimulus to another. They are moving their narrowly focused spotlight back and forth from one spot on the stage to another. While their focus is on the one stimulus, they are scanning the other with their system one for anything to alert them to divert their system two back to the other stimulus. Jayden isn't really listening to a book-on-tape while reading another on his e-reader. He's listening to the book-on-tape and processing the words coming through his oversized headphones while his eyes glaze over as they mindlessly follow the words on the e-reader. If you were to stop him at this moment and ask him what he was reading on his e-reader, more than likely he wouldn't be able to tell you. This means that students gazing intently into their laps during the teacher's lesson are not internalizing any of the information because it's only being scanned with their system one.

## If You Think You're a Multitasker, You're Not

Multiswitching is an extremely inefficient mode of mental processing. A study done by the Brain, Cognition, and Action Laboratory at the University of Michigan showed that multiswitching between two different activities not only doesn't save time, it significantly increases the time it takes to complete each individual task. Although the amount of time varies depending on the complexity of the tasks being performed, the researchers Joshua Rubinstein, David Meyer, and Jeffrey Evans estimate that time wasted can be as high as 40 percent of the time spent multitasking. This can be attributed to the time it takes to physically adjust the multitasker's focus as well as the time it takes the brain to get reacclimated to the different task.

It gets worse for the digiLearner. According to a study done by the University of Utah, rather than being better at multitasking, people who frequently do more than one activity at a time (as we know, most young people) are actually worse at it. According to one of the authors of the study, Professor David Strayer, "The people who multitask the most tend to be impulsive, sensation-seeking, overconfident of their multitasking abilities, and they tend to be less capable of multitasking." Because frequent multitaskers are so used to shifting their focus between different stimuli, they lack any real ability to truly focus on any one task.

Amid all this talk of students being overworked, the 2010 Kaiser survey that found that average elementary-age students were spending 7.5 hours a day on entertainment media also found they spend a meager sixteen minutes a day using their technology for doing homework. The nine hours a day the Common Sense census figures the average young person is on media excludes time spent on technology for the purpose of schoolwork. Between the eight hours a day a young person spends at school, the nine hours she spends on her technology, and the seven hours she spends sleeping, there doesn't seem to be too much time left in the day for homework.

This digital distraction masquerading as multitasking isn't just affecting homework. Years ago, I would give my class of ninth-grade English language learners the assignment of writing a five-paragraph essay based on several pages of information they had to read and interpret. After about an hour of class time, 90 percent of the students would turn in a completed paper, despite needing extra time to translate phrases from their native language to English. Today, I give this exact same assignment to my ninth-grade honors students. Unlike my previous students, all of these digiLearners are proficient in English. Despite this, instead of an hour, I give students three class blocks amounting to four and a half hours of time to complete it. That's just time spent in class. They are also free to work on this assignment at home. This year, when the four and a half hours were up and it was time to turn it in, 17 percent of the students hadn't finished and 8 percent of students hadn't even started it. These honors students chose to spend four and a half hours playing on their cell phones and take a zero on a major assignment rather than spend one minute working on it. Parkinson's law states that "work expands so as to fill the time available for its completion." This principle is operating on steroids among today's youth.

Modern technology's limitless potential is also its biggest problem. Laptops, phones, and tablets can do too many things, making them poisonous for the easily distracted mind. Recently, I had my students complete an essay in class. It was a "timed writing" modeled after the document-based questions their AP test would contain. Just as they would during the actual AP exam, I had my students write their essay by hand. However, three of the students had accommodations requiring they be given access to an electronic word processor for any in-class written assignment. When their hour and a half was up, every student had finished, with the exception of the three with laptops. These three had spent their time looking up related facts on Google, checking their e-mail, and looking at social media. One of the three hadn't even begun writing his essay.

This emphasizes the importance of holding firm to deadlines in both the classroom and at home. The movement in education today is to allow extensions on all assignments and eliminate penalties for not getting them in on time. Many school districts have eliminated the use of 0 percent as a grade. Students now receive partial credit on assignments they never even attempted or turned in. Many policy makers believe receiving no credit for assignments they didn't do is too harsh a penalty. Besides being counterintuitive (if modern students were in fact much better at multitasking, they would be better at getting their assignments completed on time and policies like these wouldn't be necessary), it is also counter-productive. Students who are perpetually distracted struggle to get things done in a timely manner. The more we shield students from the consequences of their inability to focus and complete their work, the more they will continue to struggle. Together, parents and teachers need to allow consequences rather than continuing to enable these unproductive habits.

One of the great-sounding selling points of the educational-technology movement is that a completely digital classroom allows students to "work at their own pace." Without deadlines students can be more free to learn when they see fit, and take the time they need to learn. But again, one must wonder if technology has made things more efficient, then why is asking digital natives to meet deadlines becoming increasingly problematic? However, the reality is modern technology has not made things more efficient. Because of the propensity for distraction, it simply takes kids longer to do things when they're digitized. Allowing students to "work at their own pace" is throwing in the towel and giving in to the fact that kids just can't focus and be as productive as they used to be. Rather than continuing to encourage focus, schools are now diminishing their standards and lowering their expectations to accommodate the new status quo.

Another reason multiswitching is detrimental to one's ability to process information is due to what professor Clifford Nass referred to as an "irrelevancy problem." Imagine tuning in to a movie that you are totally unfamiliar with halfway through.

At first you would be unable to distinguish the main characters from the insignificant extras. You would be unable to distinguish between the important and the trivial. Now compound this confusion by flipping back and forth from that channel to another channel showing something completely different. This is exactly how the digiLearner goes through life, constantly coming into discussions and lessons halfway through with little to no context. They therefore can be clueless, unable to decide what is important and what isn't. Lacking all context, they are completely unable to put this information into a meaningful narrative. What they do hear seems to lack importance. The brain tends to discard "random" information as meaningless.

Nass, the late Stanford sociologist, and fellow researchers Eyal Ophir and Anthony Wagner did a study in 2009 on our ability to multitask. Their work sought to test the claim that people who regularly multitask are actually better at it than people who don't. His research divided subjects into two groups, "chronically heavy and light media multitaskers." They did a series of three tests. The first test was aimed at testing focus through filtering out distractions. They did this by flashing two images of two red rectangles. The subjects had to determine whether the two red rectangles were in different positions in the two images shown. Both the heavy multitaskers and the light multitaskers were able to complete this task with virtually identical levels of accuracy. However, when the researchers added more irrelevant information (in the form of blue triangles) the heavy multitaskers did much worse than the light multitaskers. The heavy multitaskers had problems focusing on the relevant information while discarding the irrelevant information.

The second test sought to gauge multitaskers' ability to remember information they were exposed to. Maybe they are unable to filter out irrelevant information but have an above-average ability to remember what they see and organize their memories in a more meaningful way. In this test, subjects were exposed to a seemingly random sequence of letters and would have to indicate when a letter was repeated. Unfortunately for the chronic multitaskers, their performance in this test was once again significantly worse compared to light multitaskers.

The final test sought to gauge the speed at which a chronic multitasker could switch from one activity to another. The subjects were shown a series of numbers and letters. At one point, they were asked to focus on the numbers and indicate whether they were odd or even. Then, they were asked to switch their focus from the numbers to the letters and indicate whether they were vowels or consonants. Yet again, the light multitaskers outperformed the heavy multitaskers.

Professor Nass best summed up his findings in an interview for NPR: "People who chronically multitask show an enormous range of deficits. . . . They can't manage a working memory. They're chronically distracted. They initiate much larger parts of their brain that are irrelevant to the task at hand. And even—they're even terrible at multitasking. So they're pretty much mental wrecks," Nass concluded. He couldn't have been more right. These are our kids.

## Multiswitching: Bad for One's Health

Besides being "suckers for irrelevancy," having poor memories, and being chronically distracted, multitaskers have several more mental shortcomings. A study by the University of California, Irvine, showed that the act of multitasking increased the multitasker's heart rate and levels of stress. Chronic multitaskers experience brain atrophy and, as a result, literally have smaller brains. A 2014 study showed frequent multitaskers had decreased gray matter in the region of their brain responsible for processing information and emotions, making them more impulsive and less capable of controlling their behavior. And a study done by the University of London found that "workers distracted by e-mail and phone calls suffer a fall in IQ more than twice that found in marijuana smokers." If you've ever had a conversation with a habitual marijuana smoker, you realize how significant this finding is.

Putting these cognitive shortcomings aside, what Nass says he found most troubling was the delusional way in which multitaskers thought they had the ability to focus when needed. A study conducted by psychologists Sana, Weston, and Cepeda and published in 2012 illustrated how deep this ability to distract the modern generation of multitaskers goes. They conducted an experiment where they gave a lecture in a classroom. The students sitting in the middle of the classroom were told to use their laptops during the lecture. The students sitting directly in the front of these laptop users as well as directly behind were instructed to keep their laptops away and focus on the lecture. The laptop users were asked to take notes on the lecture while completing unrelated tasks. These tasks were the kinds of things digital natives would typically do during a lecture, such as reading e-mail, browsing their Facebook feed, and looking up things on Google. At the end of the lecture, all the students were asked to turn in their notes.

By this point it shouldn't come as a surprise that the students who were asked to multitask did poorly at both taking notes and completing the unrelated tasks. Also not surprising, when given a test on the material the multitaskers scored significantly lower than the nonmultitaskers. However, what is surprising about their findings is when they compared the test results of the students sitting in the front of the multitaskers to the students sitting behind the multitaskers, the nonmultitaskers sitting in the back of the room did significantly worse than those sitting in the front. The students sitting in front of the multitaskers only had a view of the professor. The students sitting behind the multitaskers not only had a full view of the professor but also had a full view of the laptop screens of the multitaskers. The laptop screens created a "cone of distraction" where they not only distracted the laptop users but also distracted the people sitting behind them. This illustrates a kind of "secondhand smoke" effect of multitasking. A student who makes a conscious effort to focus during a lesson may still suffer the same effects as less disciplined students sitting around him. As many class sizes increase, students sit in closer proximity to each other. After years of budget cuts, many schools have to cram thirty or more students into a room designed to fit twenty-five. Students are practically sitting on top of each other. Because of that, students can often see and hear what the student sitting next to them is doing. It has become easier to spot a kid who's playing a game or watching a movie on his phone because normally he'll have the two students sitting next to him also staring intently into his lap.

Unfortunately, the ill effects created by an entire childhood of multitasking may be irreversible, severely impairing one's ability to focus as an adult. The brain is more "plastic" than "elastic," meaning that it's malleable and adaptable in the developmental stages of childhood. However, once hardened it becomes more difficult to change as an adult. Heavy multitaskers have trained their brains to constantly search for alternative stimuli. By doing this they neglect the regions of their brains designed to focus. Eventually, and with enough neglect, these regions of the brain wither, and die. Because of this, the multitasker cannot suddenly make the decision to put down his phone, ignore everything going on around him, and focus on a lesson—or anything for that matter.

Without the ability to focus, the digiLearner feels the "need to multitask." Like Jayden's parents said, multitasking is "how kids work today." This apparent "need to be distracted" by a device in order to focus may come from a type of mental energy-saving mode we humans have. The casual scanning of system one burns far fewer calories than the in-depth thinking of system two. Having spent much of our human history in the calorie-scarce world of the hunter-gatherer era, the body has developed what Kahneman calls the "law of least effort." This law states, "If there are several ways of achieving the same goal, people will eventually gravitate to the least demanding course of action. In the economy of action, effort is a cost."

Students have come to depend on electronics like their phones and laptops to calm their overstimulated minds. They lack the mental stamina to go through an extended period of time with high intellectual arousal, so they use their devices as a type of mental pacifier. If given the choice of listening to a thought-provoking lecture, answering critical-thinking questions, focusing on a challenging equation, or playing *Candy Crush*, they will crush candy. Their electronic devices allow them to slow down their higher-level thinking by turning their focus to a menial repetitive task with which they are more comfortable.

Educational-technology advocates claim that the digital natives' prolonged use of gaming apps has actually increased their ability to focus. After all, kids are intently staring at a screen for over seven hours a day and can play a single simple game for hours on end. Makes sense, right? However, this is not genuine focus. These games exploit impulsive and addictive behaviors (as outlined in chapter 2) rather than cultivate attention. Now, game makers are adding a slight "educational" backdrop to their model of addiction and sell it to school systems as "educational technology." They flaunt how their games increase "student engagement" (addiction) as their measure of success. They show how students are captivated by their games more than the old "boring" methods of teaching. In other words, they can get a group of kids to play games. So what?

Getting kids to stop playing games is the real challenge. In class, when I ask heavy media multitaskers to put their phones or laptops away and pay attention, they typically can only focus for several moments before putting their heads down and falling asleep. It doesn't matter what the activity is or how engaged their peers are. They will say something like, "This is too boring," and imply it would be better if the lesson were in an electronic form. When protechnology advocates say something like, "It's the teacher's job to engage the students in a way that's meaningful for this generation of students," they're really saying, "Teachers need to entertain kids so they're not bored." The lessons they propose play into digiLearners' inability to focus. They rely on constant topic changes, interactive games, students working at their own pace, and distractions in the form of videos and random noises. However, the issue for students isn't boredom. The problem is that the act of focusing requires too much energy for their untrained minds, so they feel an overwhelming sense of exhaustion any time they're asked to concentrate. If schools don't teach young people to focus, who will?

## The Need for Edutainment

As teachers, we want to make every lesson as fun and engaging as the games they play. We dream of lessons in which the entire class hangs on our every word, asks thought-provoking, relevant questions, and gives us a standing ovation when we're through. There are some teachers who almost make this a reality. They have a gift for making the subject matter come alive. Unfortunately, even for the greatest teacher, this isn't always possible. Not every lesson can be a blockbuster for every student. Students' interests vary. Some students may find the topic of multivariable equations fascinating while others find it dreadfully boring. This shouldn't be the primary concern of the teacher. Teachers have to present the lesson to all students regardless of their level of "interest." But more important, some concepts aren't fun for anyone. They're difficult, complicated, and bland. Students find these concepts "boring" because making sense of them requires deep focus and repeated exposure to obtain true comprehension. In other words, learning them is hard work. Unfortunately, the concepts that kids have to work to understand are often the most important ones and the basis for entire units. Despite ed-tech firms' claims, many of these concepts can't be mastered through games or apps. Even if they could, what kind of message are we sending to our young people if they only need to focus when entertained? Will their employers have to create games to make work more fun?

These studies only scratch the surface. Despite the overwhelming amount of evidence, schools are continuing to profoundly alter their existing methodology to encourage more multitasking by means of technology in the classroom. They are replacing paper textbooks with online ones. Teachers are being encouraged to use websites to administer tests and encouraging their students to use social media during class. Some school systems are going as far as purchasing every student a laptop or iPad for personal use. Counselors are mandating that students diagnosed with ADHD be given all assignments in an electronic form. This is analogous to holding an Alcoholics Anonymous meeting in a bar. Students are now being required, in many instances, to spend their educational time in the land of distraction. This is because, like many parents, school systems don't understand the myths behind multitasking.

## Helping Your Multitasker Learn to Focus

Let's finish this chapter by returning to Jayden and seeing what we can do. Jayden's inability to focus became most apparent when the time came for him to do his homework. He was struggling through four or more hours of schoolwork a night. And even then, he was still doing poorly in many of his classes. Jayden's parents became exasperated. Like so many parents of struggling students, Jayden's parents didn't know who to blame. They knew their son was very bright, but these assignments were consuming far too much time. The logical conclusion was that his teachers were assigning too much homework. They began championing the growing movement of abolishing homework in schools. However, in Jayden's case it wasn't the amount of work that was the issue. It was the amount of time it was taking him to complete it. When Jayden's parents came to me to discuss his low test scores, the first thing they said was, "We don't understand. He spends over four hours a night working on homework for your class."

Like many parents, Jayden's didn't know a lot about the kind of work he was doing. Now that he was in high school, they encouraged him to work independently. Gone were the days of sitting with him and "holding his hand" through his assignments. All they knew was that Jayden would dutifully go to his room after dinner to start his homework for the following day. Many nights, he'd toil away until midnight or later. Jayden had signed up for honors-level courses, and his parents were concerned that he had bitten off more than he could chew. They wanted to move him out of the class because they assumed Jayden couldn't handle the workload required in these higher-level courses.

I urged Jayden and his parents not to rush to drop the course just yet. Jayden was certainly bright enough to handle it, and the workload shouldn't have been the issue. It wasn't until I explained to them what Jayden was doing for homework that they started to understand the problem. Together, we looked at a recent assignment that gave him trouble. It was a typical homework assignment for my class. He had to complete four short-answer questions based on three pages of reading from his textbook. He had two nights to complete them. It should have taken him no more than thirty minutes to do. The first question was "How did the Code of Hammurabi create social inequality?" Jayden had answered, "It was good." His remaining three responses were just as fragmented and incoherent. Now, Jayden's parents say he spent eight hours on this assignment. That means he was writing at the rate of one and a half words an hour. But what was more problematic was the fact that Jayden clearly wasn't focusing on the material in the first place.

"Either Jayden is functionally illiterate, or he's not really trying," his dad concluded. The four of us devised a plan to review Jayden's study habits so that we could come up with a more efficient and effective way for him to do his work. The first thing I asked Jayden was, "What other things are you doing when you're working on your homework?" Jayden admitted that he normally had Netflix on his television to create background noise in concert with music he was playing through his laptop. "The noise helps me focus," he claimed. Because his textbook was online, he was reading the chapter on his laptop. However, he had Facebook, Instagram, and different messengers simultaneously opened in other windows. Jayden and his parents began tracking his phone usage through an app that tallies how many times users unlock their phones and how long they spend actively on their phones. He was unlocking his phone over two hundred times and actively using his phone for over two hours of the time he was "doing homework." He admitted he would regularly respond to text and Snapchat messages. It became clear that very little of the time Jayden was dedicating to homework was actually dedicated to doing homework. Instead, he was spreading his attention between at least six different activities—and doing a poor job at each.

The solution was to create a more conducive, distraction-free, focus-friendly environment. Because most of his friends were online and trying to interact later at night (while they, too, were trying to do their homework), Jayden flipped his leisure time with his homework time. Rather than relax when he got home and do schoolwork right before bed, he would start his work soon after getting home and use the evening for relaxing, playing games, and chatting with his friends. His parents set up a "charging station" in the entrance of their home. As soon as he walked in the door, he would plug in his phone and other electronic devices and leave them there until his homework was complete. Jayden felt he would be more productive working in the view of his parents, where they could help him stay on task. He began to do his work at the kitchen table rather than in his room by himself. While he was working, the TV would remain off and he wasn't allowed to listen to music with lyrics. Although many kids claim music helps them focus, music with lyrics provides the same distraction as TV. However, nonlyrical music such as classical, or music with words in an unfamiliar language, may actually augment attention. But when given the choice of Bach or the Gipsy Kings, Jayden chose to work in silence.

The bigger challenges came when his schoolwork was to be completed electronically. The textbook for my class was a digital one, and required him to be on a laptop. But he struggled to be online without answering the siren call of social media and other digital distractions. Even when he was being productive, he found himself Googling words he didn't know, which would quickly lead him down a rabbit hole of distractions. Blinking lights, alerts, pop-ups, and push notifications all vied for his attention. This would be too much for anyone, especially someone with a young developing mind and an inability to focus. The solution for this was relatively simple. He asked me for a paper version of his book. Problem solved.

Parents and students shouldn't be afraid to ask their teachers for a paper version of their textbook as an alternative to an online one. However, you should be prepared for some pushback. When many schools switch to online books, teachers are told that the few paper textbooks they have are designated as "class sets" and cannot be sent home with students because it is prohibited by the contract with the publisher. From a teacher's perspective, however, it is very difficult not to help students who want to do their homework. Students like Jayden who are heavily tech dependent can't be expected to use an online text productively. If the teacher won't or can't provide a paper version of the book, see if it's available for checkout at the school or local libraries. If that doesn't work, ask the counselor to create an accommodation for your student, allowing him or her access to a physical book for each class.

With some assignments, the use of technology cannot and should not be avoided, such as for research or writing a paper. For these, I recommended Jayden handwrite an outline of his project or paper before picking up the computer. When doing research, I had Jayden hypothesize about what types of resources are likely to have relevant information, where he would likely find primary sources, and what Google searches he was going to use. For example, if he was doing research on the reasons for the Crusades, I would have him start by listing possible sources of relevant information. If he was looking for primary sources, I would have him list important people such as Pope Urban II, Saladin, Richard the Lionheart, other lords, bishops, or monks. Once he created this list, he would then generate some sample searches, such as "Pope Urban II speech," "Crusade monk journal," or "Crusader diary."

When writing a paper, I would have Jayden create a detailed outline by hand before he started writing. He would develop main ideas, and organize the supporting evidence for each paragraph. When it came time to write, it was just a matter of connecting the dots. He would prepare his rough draft by hand so that he was only transferring it to a computer in its final form. He was less likely to be distracted when he had a formal plan by the time he picked up the computer. These are all steps you can take at home.

The final part in creating a focus-friendly environment was having his parents model this behavior. Both Jayden's parents spent a lot of time on their laptops and phones for work. Even after their workday, they still felt tied to their devices. They would find themselves regularly checking e-mail, messaging coworkers, and, like their children, they would use their devices for games to help them unwind. They, too, became prone to distraction. They struggled to focus on a single task for an extended period. They would glance at their phone regularly or even completely check out during conversations with their family members, or even during dinners. They began to leave their phones by the door and started dedicating themselves to one task at a time. When they ate dinner with their family, it was a time completely dedicated to being with their family.

These changes were easier said than done. The distractions had become such an important part of Jayden's life that removing them was challenging. He didn't give up his phone easily. It took commitment and sternness on the part of his parents. However, the selling point with Jayden was when he saw the results for himself. It only took about a week before he started to show significant improvements. He was only spending an hour doing his homework rather than four or more. Because of this, he actually had more free time. He could now dedicate his nights to the kind of activities that were distracting him in the first place. Further, the quality of his work improved. He was now giving more insightful responses that showed a deeper understanding of the material. His new ability to focus carried over to his performance in the classroom. I noticed that he was more alert and attentive in class. And unsurprisingly, his grades and attitude improved.

## TAKEAWAYS

- Teach focus. Allot at least thirty minutes a day to focused thinking without distraction. During the school year, have your child dedicate this time to doing her homework or studying. During vacation, have your child read a book or learn about something he has always been curious about. It should be something that's interesting to your child but that doesn't provide instant gratification like a video game.
- Lay down the law and don't let kids persuade you. Remember, focus requires effort. It's hard work. Naturally, kids will try to avoid it. They will use arguments like, "That's not my learning style," or, "I'm better when I'm multitasking," or, "Background noise helps me concentrate." They use these arguments not because they're true, but because they work. Don't fall for them. If the adults in kids' lives cave in and give them what they want (their phones, laptops, video games), that only reinforces their misguided notions. The ability to focus is an incredibly important life skill. Like eating their vegetables, it's not something kids want to do at first. But because you understand its importance, you make them do it anyway.

- Take kids' devices away when you expect them to con-centrate. Electronic devices have no place anywhere near spaces where focus is required. Don't allow kids to have them in class, and don't allow them to have them when they're doing homework. Not only will your child be better off, but, as we have seen, so will the students sitting around him or her.

# Escaping the Digital World of Anxiety

*ASIANS, DON'T KILL YOURSELVES*
—SCHOOL SIGN FOR SUICIDE PREVENTION MONTH

SIX OUT OF FIFTEEN. In percentage terms, it's exactly 40 percent. It's also the number of girls that were hospitalized for depression, anxiety, or attempted suicide in one teacher's class. This colleague from a different school, I'll call her Dana, reached out to me after this devastating and tragic school year.

As any teacher will tell you, classes can take on personalities of their own. Like a flock of starlings, a group of thirty kids can behave and act in a singular fashion with seemingly no rhyme or reason. You can have a bunch of normally well-behaved individuals make up a horribly behaved class, typically entertaining kids make up a boring class, or a group of boring kids make up an exciting class. In Dana's case, the moment she walked into the room this particular year, she could feel the tension in the air. It's not that these kids were "bad" or poorly behaved in a traditional sense. The painful part of being their teacher was their suffocating negativity. Many of them, especially the girls, were always in an uproar about something, normally involving something said by a boyfriend or girlfriend, a friend, a parent, or a teacher. What Dana came to learn the hard way was that these girls, who were prone to group drama and negativity, were actually exhibiting signs of depression and anxiety.

Staci was the first student to make this clear. One of the class's alpha females, she had a strong, dominating personality and ruled over her clique with an iron fist. She felt resentment toward authority and strongly disliked being told what to do. She struggled in school and engaged in a daily fight with Dana over Staci's perpetual use of her phone, hair brushing, and talking to her girlfriends throughout class. Dana has no counseling or psychological credentials, and she assumed Staci was just severely narcissistic. Shortly after the first semester, Dana was caught off guard when she received news that Staci wouldn't be returning to her class. Staci had tried to kill herself. Dana learned she had been cutting herself with a razor blade in the restroom, and eventually tried to slit her wrists. Although the resulting wounds were mostly superficial, her underlying problems clearly ran much deeper. She spent the remainder of the school year in a mental health facility.

Dana barely had time to process what had happened with Staci when a week later another one of her students was also hospitalized for attempted suicide. Riley was a quiet, reserved, and conscientious student. She stayed away from the drama that surrounded Staci and her friends. So it was surprising to learn she had taken a bottle of pills after an incident involving her boyfriend, who had recently broken up with her and posted ugly rumors about her on social media.

After that, it seemed like every month Dana had another girl exhibit mental health problems. Jules had run away from home because she was wanted by the police for assault. She had repeatedly attacked another girl because of an ongoing cyber-bullying incident. When the police found her, they determined she was too unstable to return to school. She spent the remainder of the year in a school for emotionally disabled teens. Then there was Eileen, who was hospitalized for anorexia. And finally, there was Aria, who suffered a concussion while playing soccer. Severe concussions can make serious thought and concentration excruciatingly painful. As a result, students diagnosed with one are understandably excused from all classwork and homework, encouraged to stay at home and sleep, and, above all, stay away from digital screens. Aria's concussion lingered for months (mostly because she couldn't seem to follow the doctor's order to stay away from screens) and she had consequently fallen far behind in school. Eventually, her mom appealed to her counselor. Aria was suffering from depression and anxiety as a result of her concussion and mounting makeup work. She wouldn't be finishing the year, was excused from all her missing work, final exams, and standardized tests, and her final grade was to be based on her preconcussion grades. These girls were struggling in numbers Dana had not seen or dealt with before.

## Greater Awareness Yet Declining Mental Health

Luckily, a class like Dana's is not the norm. In years past I would expect maybe one or two of my 150 students to be dealing with some type of mental health issue. It's a sad reality when working with teenagers. However, this number is growing. Now, there are normally several issues per class. It's an alarming trend among digital natives across the country. Deaths by suicide have more than tripled in the last decade among adolescent teenaged girls, and are up 60 percent worldwide in the last fifty years. It has become the third leading cause of death among teens. The annual rate of suicide has doubled since 2006. And these statistics don't even include attempted suicides like Staci's and Riley's. This is just one indication of the drastic decline in the mental health of young people.

Trouble dealing with anxiety is quickly becoming a defining characteristic of this modern generation. A San Diego State University study compared seven million adults and adolescents from 1980 and 2010. The study found that, when compared to digital immigrants, digital natives are 38 percent more likely to have trouble with their memory, 74 percent more likely to have trouble sleeping, and college students were 50 percent more likely to report feeling overwhelmed by both their academic and personal demands. Today, one out of ten young people reports having major issues with depression.

As a result, schools have understandably become extremely sensitive to the growing problem of teenage suicide and depression. Like many school systems, mine has taken extreme measures to help ensure no kid is made to feel depressed as a result of his or her experience at school. As many others are beginning to do, the start of our school day has been moved back one hour to give kids more time to sleep in. Lack of sleep has been linked to several major issues relating to poor mental health. To lessen the pressures students and parents put on themselves, grading scales in many school systems are being made less rigorous. Many are also doing away with valedictorians and class ranks because they add too much pressure and competition on kids to be number one. Many have also done away with rigid mandatory punishments, allowing more flexibility when dealing with students who may have extenuating circumstances. Most schools now have some sort of suicide prevention week or month. Signs adorn hallways with slogans like ASIANS, DON'T KILL YOURSELVES followed by a statistic about suicide rates among Asian American teens. As teachers, we were once encouraged to grade only in green ink and to mark only correct answers. This suggestion came from a school survey in which students labeled their teachers as one of the top bullies in the school because we were always "pointing out their mistakes." The point is, schools are trying desperately to do almost anything to help this dire problem.

Many of these changes came on the heels of the documentary *Race to Nowhere* and similarly themed books and articles that started to appear in the early 2000s. This movement brought much-needed awareness to the issue of mental health in our teens. However, much of this work attributes declining mental health to the increasing pressure on young people placed on them by their schools. The increasing rigor of school, amount of homework, pressure to participate in extracurricular activities, and competition of getting into college were all proving to be too much for the modern young person.

However, none of these reasons seem to be supported by what's actually going on in schools. First of all, if school was becoming more difficult, you would expect average grades to be dropping. But that's not the trend. According to the US Department of Education, since the '90s, the average high school GPA has risen 0.33 on a 4.0 scale (that translates to nearly half a letter grade) while student scores on standardized tests like the SAT and the ACT have remained fairly constant. This would indicate that although students today have a similar level of knowledge when compared to their digital immigrant counterparts, they have significantly higher grades. Much of this grade inflation can be attributed to an increasing number of schools switching from a seven-point grade scale (where an A is 93–100 percent) to a ten-point scale (where an A is now a 90–100 percent), which improves letter grades (and hence, GPAs) without improving student percentages or performance. If schools were in fact becoming more difficult, you would expect to see GPAs decline, or remain the same while scores on standardized tests went up. You certainly wouldn't expect to see grades improve.

There also doesn't seem to be much validity to the claim that students are expected to do more work in the modern era of education. Homework rates have remained relatively consistent over the last thirty years, the same time period that suicide rates have skyrocketed. In a series of surveys conducted by the National Assessment of Educational Progress that asked students how much homework they had the previous night, both in 1984 and 2010, 26 percent of thirteen-year-old students admitted to doing no homework. Seventeen-year-olds have fared a little better, with 38 percent claiming they did no homework in 2010 compared to 33 percent in 1984. The number of students claiming to have done over two hours of homework per night declined 2 percent for thirteen-year-olds from 1984 to 2010 (going from 9 to 7 percent) and for seventeen-year-olds, the percentage remained the same: 13 percent. Sixty percent of students are averaging a little over forty minutes or less a night on homework.

And this stress isn't coming from an increase in participation in extracurricular activities, either. The percentage of high school students who have jobs dropped by half in the time frame in which suicide rates tripled, going from 32 percent in 1990 to a record low 16 percent in 2012. This is while participation in extracurricular activities has remained around 60 percent since the 1970s. Reports of young people being overbooked with after-school activities may be exaggerated. A recent Pew Research Center survey found that while only 15 percent of parents complain their child's day is "too hectic," nearly half of parents claim their children "get too much screen time."

The narrative that modern children's declining mental health is mainly caused by today's academic pressures doesn't fully explain what teachers are seeing in the classroom. Like most of the girls in Dana's class, a majority of the incidents involving suicide and depression were rooted more in social pressures, such as problems with friends, girlfriends and boyfriends, body image, and low self-esteem. For most of these students, their interest in education was very low. Looking at the bigger picture, when you consider that 64 percent of seventeen-year-olds are spending little to no time at night doing homework but logging nine or more hours on their screens, it becomes clearer where the problem lies.

Psychologists say this has been an obvious conclusion for some time. There have been dozens of studies showing a strong correlation between excessive technology use and a decline in mental health. One such study done by the Kaiser Family Foundation found that "heavy digital users" are more likely to be sad and unhappy, bored, and much more likely to get into trouble. A study in Scotland found that total screen time alone was an accurate predictor of psychological distress. Another in Australia found that young people who spent excessive time on their screens had a much higher rate of feelings of "loneliness, depression, withdrawal, anxiety, attention problems, and aggression." A 2011 study found that teens spending excessive amounts of time on screens experienced higher rates of "sadness, suicidal ideation and suicide planning." A 2014 study found that excessive technology use was a likely factor in "poorer health that is defined as psychological issues, behavior problems, attention problems, or physical problems." The study goes on to say that although researchers found that overuse of video games, "technological toys," and "electronic communication" were particularly harmful to preteens, "overuse of any technology" appears to have a negative impact on teens. Another study found that young people started experiencing many of these symptoms of psychological distress after just one to two hours of screen time a day. Considering the average time on screens is closer to nine hours, these studies describe an entire generation of kids. How is it possible that these devices could lead to such harm?

## The Digital Pit of Despair

One of the greatest paradoxes of the digital age is the digital native's increased feeling of loneliness. With all the social media and web-based interaction in use today, people can stay in touch with a much larger network of people all over the world. This is one of technology's biggest selling points for educational markets. Social media and networking allow for students to gain access and share information with experts and other people all over the world, making students more "globally aware." Despite this increasing ability to be constantly interacting with others, young people today are reporting greater feelings of withdrawal and loneliness. Perhaps most surprising is that it is those teens who spend the most time on social media who feel this most acutely. This leads to an obvious conclusion: digital interactions are not the same as face-to-face ones. People don't have the same positive reaction to a positive online interaction as they do to in-person ones. Our need to "fit in" and be accepted by our peers is more than a recent fad. This is an element of humanity that is deeply woven into our fiber. Locking oneself in a room to be on social media is not an adequate substitute for real-life relationships.

A student I had several years ago proved this. Like most kids his age, Brayden was a little awkward and painfully shy. He was an extremely nice young man but didn't seem to have many friends. He was in my class for the first period of the day, and every morning he would arrive to my room thirty minutes early. He would sit in his seat, silently glued to his cell phone screen waiting for class to start. He voluntarily chose this solitary start to his day rather than face the masses of students congregated in the hallway. As the year went on, I became worried about Brayden. He began to demonstrate many of the warning signs of depression. I reached out to his mother, and she confirmed that he was expressing anxieties about not fitting in.

After spending time working with him, I could tell that he really wanted to come out of his shell and forge relationships with some of his peers. He was a good-looking kid, but would reinvent his image every month. He would look longingly at the other students who were talking with their friends in class. But he could never take the leap and actually join others in conversation. Instead, any time there was an opportunity for conversation, he would immediately pull out his phone and put in his earbuds. He used it like a security blanket, something he would clutch whenever he felt uncomfortable. And it became obvious he felt uncomfortable in most social settings.

One day, I gave an assignment that required small-group discussions. As I was putting it together, I couldn't help but feel uneasy about Brayden. I worried about how he would handle the setting, and if it would cause him too much anxiety. This is in part why I decided to do something I don't normally do, assign everyone to a random group. This way, I could ensure Brayden wasn't left out.

Each group was given a story about a lonely young high school student who punches a popular jock in the face for reasons only known to the lonely young man. It was an allegory for the events of 9/11. After reading it, the students were asked to have a conversation about what the jock (and the United States) should do next. This was not a graded assignment. The point was to get the students talking to each other about foreign policy through the medium of the high school drama they were more accustomed to.

Brayden loved it. For the first time, he was engaged, talkative, and excited. As I walked by his group, he was giving them an interesting perspective on the lonely young man that the other members of his group clearly hadn't considered before. He emerged from his shell and was even downright charming at times. After class, he approached me and said, "Can we do group projects every day? Group projects fit my learning style."

First of all, "group projects" are not a learning style—even if you believe in the theory of different learning styles (there is some new research discrediting this idea). Regardless, I could see what he was saying. The following week, I gave the students another opportunity to work with each other. This time, however, I told them to get into their own groups. And because I was grading their work, I gave them the option to work alone (this is a policy encouraged by our county when assigning graded group work). Much to my surprise, Brayden opted out of working with a group, choosing instead to sit in the corner by himself. When I approached him to ask why, I could tell he was seething with anger. He sat there, clenching his pencil in his fist with a scowl on his face that would have frightened Bill Cowher.

"What's wrong, Brayden?" I asked.

"No one wants to work with me," Brayden replied through his clenched teeth.

"Did you ask anyone to work with you?" I asked, even though I knew he hadn't. I had wanted to see him continue to emerge from his shell but watched disappointedly as he sat there, glancing downward at his phone to avoid eye contact with the other students as they looked for partners.

Fighting back tears, Brayden explained, "You need to assign the groups—that's how I work best."

Brayden reminded me of the famed studies done by Harry Harlow on the nature of love. Harlow became infamous for isolating baby monkeys in order to prove physical contact and social interactions were a basic necessity of life. One of his most notorious experiments involved putting baby monkeys into a cage-like structure he called "the pit of despair." Even though the monkeys were given food and water, they were denied contact with all other animals and people. Unsurprisingly, the isolated monkeys all developed severe, and often irreversible, psychological disorders. Like Harlow's monkeys, Brayden spent much of his day isolated from others. His mom confided that he spent most nights locked in his room on his computer. She said he was typically in chat rooms or gaming with "friends." But the people he was interacting with he had never met and only knew through a digital means. He was developing superficial bonds that left him unsatisfied. So, for Brayden, school was it for him. It was his only chance to fulfill his inherent need to be a social animal and have real interactions with others. When this need went unfulfilled, it caused him a great deal of depression and anxiety.

I worry about how kids like Brayden will respond as schools become ever more digitized. Both their time at home and at school will be spent in front of screens. With no real contact with others, they will come to live in their own digital pit of despair.

With so many kids deficient in their fundamental need for real human interactions, many are starting off with deep underlying depression and other psychological problems. Rather than helping, social media only exacerbates this problem. It becomes a place where damaged kids can spread their fears and worries to their peers.

## The New World of College Application Pressure

One of the most pervasive of teens' fears is the misconception that it's becoming more difficult to get in to college. It is true that acceptance rates at many of the country's top universities are declining. Many of them are reporting record-low admission rates. In 2016 Harvard accepted a mere 5.2 percent of applicants, and Stanford reported a record low 4.7 percent. Many more of the top universities had admission rates of 10 percent or less. However, these statistics are somewhat misleading. According to Jim Hull, a senior policy analyst at the National School Boards Association's Center for Public Education, the idea that it's more difficult to get in to a college today than it was in the '90s or earlier is an "urban legend." His group analyzed SAT scores and GPAs spanning the 1990s and 2000s and found that the same caliber students were getting into the same caliber universities before 2000 as were after.

This "competition" is largely an illusion. The pool of applicants to college isn't increasing as fast as the number of applications. Before the 2000s, kids typically only applied to three or four schools. This is because application fees were expensive and each college had its own unique application for prospective students to fill out. Students were having to write an average of three or four unique essays per application. Today, most universities use a software called Common App, which has streamlined the application process and created one uniform application for most schools. Now, students just have to go to one website and fill out one application. Then they can send it out to as many schools as they want with the click of a button. Because of the increased ease of applying, the average student today submits more than ten applications. Tripling the number of applications would account for a decline in acceptance rates.

Another reason for the decline in acceptance rates is that a disproportionate number of these additional applications are going to students' "reach" schools (universities to which there is a high likelihood of rejection). Students have always wanted to get in to the "best" colleges and universities. This isn't something unique to this generation. What is changing is the pressure students are putting on themselves to get in to the most selective schools. As high schools across the country have moved away from ranking their students and naming valedictorians, getting into a school with a really low acceptance rate would give them recognition as an "elite student."

Last year, after one of my students returned from an extended absence, I asked him, "Where've you been, Ronnie? Vacation?" Ronnie was a Korean American student who typically went back to Korea two or three times a year.

"I had my college interview with Penn," he replied.

Ronnie was a nice kid. But he was a below average student. He was a low C / high D student who was probably my second-lowest performing student that year. He had almost no extracurricular activities, and his SAT scores were pretty mediocre. Yet here he was applying to one of the top schools in the country.

Of course I didn't say any of this to him. But he could sense that he needed a follow-up explanation. "My mom wanted to tell her friends that her son was applying to Penn, so she made me apply. I know I don't have a chance of getting in. It's a Korean thing."

This was the first I had ever heard of this, so I asked the class how many students had done something similar. A third of the class raised their hands. Apparently, it wasn't just a "Korean thing." That night when I was on Facebook, I noticed a friend of mine had posted a picture of her daughter typing on a laptop with a caption, "Application time—this one is going to Princeton." I knew her daughter well enough to know that she was a struggling C student. She didn't get in to Princeton. She got in to Radford University (not a bad school, but in terms of academic reputation, it is light-years away from Princeton).

Rather than focusing on finding a school best suited to their abilities, these students and their parents were using the college application process as a chance to present an exaggerated illusion of their own academic success. Social media has created a platform for this. Before, discussions about what schools a student was applying to were reserved for a handful of close friends and family, the kind of people who knew the child well enough to know which schools were in his reach.

Social media has brought out an unhealthy competitive side in people. It starts with misleading statements, which aren't necessarily false, like, "My son is applying to Penn." This sounds impressive in a traditional context, but taken at face value really isn't an accomplishment at all. Even farther from the truth is another increasingly common throwaway line, "I'm considering going to Princeton." One can close their eyes and consider what it would be like to go anywhere, including Ivy League schools. That doesn't mean you actually have a chance at getting in. These are the kind of posts becoming increasingly common from both children and their parents.

This facade has degraded the intrinsic rewards one experiences for actual achievements. When real accomplishments, like getting into Radford, pale in comparison to the fictional reality presented on social media, like potentially getting into Princeton, life becomes a place of perpetual disappointment. Because of this, digital natives tend to appreciate their own triumphs less. That young lady should have been proud to have gotten in to Radford, rather than lamenting not getting in to Princeton. It's as though any achievement is a disappointment unless it can be packaged and presented on social media.

It's as though digital natives (and now their parents) only seek out accomplishments that are tailor made for social media. Achievements are meaningless if they can't be photographed and put on Facebook, Instagram, and Twitter for everyone to see and marvel at. Striving for only extrinsic success by dressing up all their accomplishments to become something more than they are must leave these young people feeling hollow and empty inside. They present an image of themselves to the world that they know to be untrue. Imagine the feeling of despair they must feel when they can't live up to their own digital persona.

In 2015 a student at one of the country's top magnet high schools made international headlines when she took her fake digital persona to the next level. This high school senior announced on social media that she had reached the pinnacle of educational success. She was such a great student that both Stanford and Harvard were fighting over her. Both schools were offering her five-figure scholarships and having professors, deans, and alumni actively recruit her. "Genius Girl" (as she was dubbed in the news) even had Mark Zuckerberg, founder of Facebook, personally call her to convince her to go to his alma mater, Harvard. Unable to decide between the two, Harvard and Stanford did something they had never done before. They settled on sharing her by creating a new program just for her, one that would allow her to have dual enrollment at both schools. News outlets all over the world picked up this remarkable story. There was only one problem: none of this was true.

Genius Girl was a very bright young lady, but she hadn't gotten in to either Harvard or Stanford. She had manufactured fake admission letters on her computer, created fake e-mail addresses from fictitious Harvard and Stanford faculty and staff, and then promoted her own story through social media. When her own self-promotion turned viral, people all over the world were touting her fake accomplishments. It wasn't until several of her classmates got wind of the story and began to question it that the truth came out. What's most interesting about this story is how easy it was for her to create this imaginary world.

## Digital-World Personae, Real-World Problems

Behavior like Genius Girl's is becoming the focus of many different psychological studies. One such study, done in Germany in 2015, examined the self-images of both video game and social media addicts. When compared to a healthy non-tech-addicted control group, the video game addicts identified more with their fake gaming avatars than their actual selves. Video game addicts were much more likely to reject their self-images and have lower feelings of self-worth. To them, experiencing a world of make-believe through their avatars was more appealing than experiencing life on their own. Similarly, social media addicts had a much less rewarding experience when asked to self-reflect than the control group experienced. The social media addicts preferred to think about an "ideal self"—an image of who they wanted to be. Social media is nothing more than a presentation of one's ideal self.

In both gaming avatars and social media profiles, people can create and present themselves any way they chose. A video game addict who spends nine hours a day locked in a dark room in a sedentary position can go from being a doughy-soft, pale child to a tall, dark, muscular hero. On social media, one can make only flattering pictures and information available to others. Responses can be well thought out, and posted in their own time. One can even go as far as Genius Girl and manufacture a fraudulent reality. One can take it one step further and make one's entire existence fake, down to using pictures of someone else as one's own. This trend is so common it's been given its own term: catfishing.

For people with lower self-esteem, these fake personas can become very appealing. The more time they spend on their computers, the more withdrawn they become from the real world, the lower their real-life self-esteem becomes. Compounded by the unhealthy nature their addiction has on their physical appearance and relationships, they want to leave the real world and immerse themselves in their fake personae. The user becomes trapped in a downward spiral that must feel inescapable. Although Genius Girl may not have been the supergenius she claimed, she still was a very intelligent young woman attending one of the highest-performing magnet schools in the country. She certainly was smart enough to realize that at some point her family and friends would come to realize she hadn't gotten in to either Harvard or Stanford. She also had to know that the more real-world attention she had on her lies, the more likely she would be found out. Yet she continued to give interviews to different news media. This goes to show just how powerful this escape from reality can be for anyone. She was a successful young lady by almost any account, yet she couldn't compete with her own ideal self. This battle of real self versus fake ideal self became so intense that she could probably no longer separate fantasy from reality.

It's these ideal and fantastic worlds that become means to escape the pressures and problems of the real world. When Genius Girl didn't achieve her goal of being a top-ranked Ivy League student, she must have felt like she was a failure. Rather than coming to grips with these difficult feelings, coping with them, realizing that everything was still OK, and moving on, she chose to create and live in an imaginary world where she could postpone, or avoid confronting, her failures altogether.

Unfortunately, this escape is only temporary. At some point digital natives like her will have to confront these problems. And by that time, they will have piled up and will be all the more overwhelming. A child who resorts to a weeklong marathon of *World of Warcraft* because he's too stressed about the amount of schoolwork he has will find himself even further behind. Rather than doing a little work over several nights, he now has to do hours of work in an evening. A child who creates an avatar to avoid dealing with his own emotions will eventually find himself dealing with even deeper depression.

This avoidance is likely to lead to the underdevelopment of the region of the brain intended to handle these strong emotions. Several studies show that children who use technology for more than five hours a day (remembering that the average child is now using nine hours of technology) showed significant pruning in the regions of the brain responsible for impulse control. Besides increased impulsivity, a 2014 study found that, like pathological gamers, heavy social media users also exhibited "poor emotion regulation skills, including lack of acceptance of emotional responses, [and] limited access to emotion regulation strategies." This is a recipe for disaster. The inability to deal with one's emotions in a healthy way combined with an inability to control one's impulses could lead to extreme and dangerous behavior, such as suicide. This lack of control is why heavy tech users are more frequently getting into trouble at both school and home.

It's this increased impulsivity that makes the digital native so wildly unpredictable. While on one hand, technology can create a digital world young people can use to avoid dealing with their real-world problems, other types of technology can make their real-world problems inescapable. The speed at which they can get information gives their delicate minds little time to process it. They can check their phones for their grade moments after they finish a test. Teachers can send them reminder messages about upcoming assignments. They can download entire lessons off Google Classroom. During snow days, teachers can send out invites to watch their recorded lessons. Always being connected has allowed the pressures of school to encroach on their mental sanctuaries. Only when students and schools learn to balance connectedness with common sense will schools be the truly safe places they are supposed to be.

## TAKEAWAYS

- Have face-to-face conversations with your children. Remember that a "How u feelin??" text message is nowhere near as comforting or reassuring to a child as an actual conversation and hug. Ask questions about who their friends are, what their goals are, and how they're feeling. Our relationships with our kids needs to be the strongest constant in their lives. As psychologist Daniel Siegel and child development specialist Mary Hartzell say, "Experience shapes brain structure. Experience is biology. How we treat our children changes who they are and how they will develop. Their brains need our parental involvement. Nature needs nurture."

- Next, once again, talk to the school. Ask the teachers and principal how the school's uses of technology are helping make your child mentally and psychologically healthier. Ask what they are doing to prevent cyberbullying. Ask their opinion on how much screen time is healthy for someone your child's age. Ask if they believe the teacher-student relationship is important. Ask how screen time strengthens that relationship.

- Monitor kids' social media use. Ignore the message of ed-tech advocates that urge parents to stay away from their children's social media posts. It's not like cracking open the lock on their diary. Their posts are public, often accessible to any person with an Internet connection. Why should a parent be the only person not looking at their posts? What they choose to post can be a powerful insight into the mental state of a child. If you have cause for concern, talk to a psychiatrist. Sometimes we can misunderstand the signs of depression.

- Keep your child grounded in reality. We cannot allow our children to escape into fantastic digital worlds. If you have concerns about your child's mental health, don't get him or her "role-playing" games that allow players to assume a different identity. Even if you think your child doesn't have any issues with depression or self-esteem, put limits on social media and gaming. Don't be afraid to keep these limits throughout high school.
- Make them disconnect. Set aside portions of every night for family time to relax and be with each other. By taking children away from their digital world, you allow them to focus on reality while distancing them from the stresses and pressures of always being connected. While this might not be possible every single evening, it is a good target.

Even in high school, parents are critical to the mental health and stability of kids. We cannot surrender this role to the on-screen pursuits of a child. I provide a fuller treatment of how families can support their kids in school in the next chapter.

6

# Reestablishing
# Support from Home

*That iPad is her only babysitter.*
—An elderly man in McDonald's describing
a young girl who was left alone

Recently, while grocery shopping I ran into a woman named Susan whose three children I had taught. Her kids had all graduated from high school and she was working as a middle school assistant principal, so I was surprised to see a curly-haired two-year-old girl buckled into the front of her shopping cart. "Your youngest?" I joked. She was actually Susan's first grandchild. As Susan and I chatted and caught up, I couldn't help but notice that her granddaughter was silent, hunched forward over an iPad as she pecked at the icons on the screen in a zombie-like trance. Susan realized I was looking at the girl. "It's amazing, isn't it?" she asked, pointing to the iPad. "I remember trying to grocery shop with my kids. What a chore! Now she's occupied and learning, and I can get in and out of here in a hurry. Isn't it wonderful?"

My head nearly exploded. The only response I could come up with was some sort of guttural noise. We said our good-byes and parted ways. Rather, she moved on as I stood there, processing what she had said. Of course grocery shopping with little kids is a pain—*because they ask questions*! A grocery store is, at its core, a learning laboratory for a child. "What's this? Can I have one of these? Why is ketchup in a plastic bottle and pickles are in glass jars? Why do the lobsters have rubber bands on their claws?" Even when children are too young to ask questions, they are still aglow with curiosity. They touch everything, filling the cart with random items they've pulled from the shelf when the parent has turned away, trying to get a sense for what different things feel like. They point to different-looking people as the parent quietly pushes their hand back down. They interact with their parent and with the people and products around them. Yet Susan, who was charged with making academic acquisition decisions for her school, thought it was wonderful that she was cut off from the child, and she assumed that the girl was learning more from the tablet than she could from her own grandmother. What do you think Susan is going to say the next time Apple calls to sell her school some iPads?

## Parenting Is Hard Work

Look, I get it. Giving screens to kids makes things easier. Who doesn't like easier? I would be lying if I said I didn't at least occasionally fall prey to this temptation. However, as is the case with most temptations, there is a high cost to pay when we give in too often. One claim by ed-tech firms that rings true is that classroom behavior becomes easier to manage when kids have devices. Absolutely. That's because kids become passive when they're on screens. The room is silent. Everyone is in his or her seat. It can make life easier for teachers. The same is true for parenting. Imagine how much more work you could get done or how much more "me time" you would have if you didn't have to deal with all those pesky questions from your children when you got home from work. For generations, parents have been searching for *that one toy* that will keep kids engaged and quiet for hours at a time. Well, we now have it: screen time. Screens may be worse for kids and their brains, but they're easier for moms and dads. That's the danger. In fact, one study found that the average US teenager now spends less than thirty minutes *per week* talking with his or her father.

It used to be almost a cliché that parents are a child's first and best teachers. However, it's true. Yet today, far too often we outsource this job to screens. As child psychologist Dr. Richard Freed explains, "Family is the most important element of children's lives—even in this world of bits and bytes—because we are human first. We can't ignore the science of attachment that shows our kids need lots of quality time with us." Tech firms would love us to turn over the reins of parenting and teaching to their apps and gadgets. An additional problem, of course, is that kids need their schools and teachers almost as much as they need their parents. Dr. Freed says, "Second in importance only to family is children's involvement with school. Nevertheless, some question the value of traditional schooling, claiming that in the digital age kids learn best through exposure to the latest gadgets."

The tragic reality, though, is that kids today are less connected than ever to their families. Researchers Gustavo Mesch and Ilan Talmud point out that the empirical evidence on the connection between Internet usage and time spent with families "shows a reduction in the time parents and youth spend together." In addition, in his excellent 2015 book *Wired Child*, Dr. Freed presents research showing that the more time teens spend watching TV or playing on screens, the less attached they feel to their parents and families. Even though many people seem to believe that screen time helps kids stay more connected to their families, the opposite is true. In fact, Microsoft conducted a poll and found that 64 percent of parents aged twenty-two to forty believe that digital technologies bring their family closer together. However, that perception is not reality. A study by the *Archives of Pediatrics and Adolescent Medicine* has found that teens who spend an average or greater than average amount of time on screens per day are less attached to their families than teens who spend less time on screens. How does this affect education? The degraded relationships between teens and parents make it tougher for kids to do their best in school. Yet that's not what ed-tech firms would have us believe.

Ed-tech firms prey on our insecurity as parents. Spend an hour watching kids' TV shows. Many of the advertisements during these shows—for toddlers through teenagers—are for digital apps, gadgets, and electronic "edutainment" games that promise to help your child succeed. Tearful parents are used to testify to the miraculous turnaround in their children. If you want your children to get ahead and reach their potential, the ads tell us, then you *must* have this app, game, website membership, gadget, or whatever. Well, what parents *don't* want their children to reach their potential? We see other parents letting their kids use tablets and doubt starts to creep in. We hear that voice, *I don't want my kid to be an idiot. I want him to get ahead and reach his potential, like all those happy, brilliant, perfectly multicultural and gender-neutral kids in the ads! I better get an account on Mykidsagenius.com* right now!

However, the most enticing part of seeing those other parents letting their kids use tablets might be that those other parents look like they're having a lot more fun than we are. They're able to talk to their spouses at the restaurant (that is, if the adults are not also on their devices). They're listening to their favorite music in the car while the kids are in the back with earbuds in, pecking away at screens. They even have a chance to catch up with their kids' former teachers as they leisurely stroll through the grocery store. So, we have the "guilty parent" factor: *I want to be sure my kid knows as much as these other kids.* And we have the "me time" factor: *It sure would be nice to just read the paper and have a cup of coffee for a half hour every once in a while.* All that leads to parents spending billions of dollars on gadgets and apps that are supposed to make their kids smarter.

How's that working out? Dr. Aric Sigman, an associate fellow of the British Psychological Society and a fellow of Britain's Royal Society of Medicine, says that when parents buy tablets for their children—even for educational use only—they're doing more harm than good. Doing this, he says, "is the very thing *impeding* the development of the abilities that parents are so eager to foster through the tablets. The ability to focus, to concentrate, to lend attention, to sense other people's attitudes and communicate with them, to build a large vocabulary—all those abilities are harmed." Victoria Prooday, an occupational therapist who specializes in working with children, parents, and teachers, points out that when children are exposed to a lot of screen time—particularly screen time that provides a ton of stimulation (such as you'd find in video games)—it gets tougher and tougher for those kids to make sense out of and pay attention to normal human interaction of the sort you'd find in school. She says, "After hours of virtual reality, processing information in a classroom becomes increasingly challenging for our kids because their brains are getting used to the high levels of stimulation that video games provide. The inability to process lower levels of stimulation leaves kids vulnerable to academic challenges."

## Children Learn What They Live

So where are kids getting the idea that it's all right to spend all that time on screen? I don't like hearing it any more than you will, but many kids get that idea from their parents. According to the Nielsen Company, the average American adult now spends nearly eleven hours per day on some type of screen. Kids see us constantly on our phones, computers, and tablets. They see us feeling the need to reply immediately to texts and e-mails—both work related and social in nature. They see us playing video games, checking our fantasy football teams, looking for deals online, finding recipes on Pinterest, and on and on. What messages could a child take from that? One possibility is this: the child loves and respects Mom and Dad, she sees Mom and Dad on their screens, she wants to emulate Mom and Dad, so she gets on her screen, too. Another (worse) possibility is that the child sees the parent seeking refuge, information, escape, and entertainment from the screen and figures that the screen is more important than she is. One response to that could be for the child to want to escape and seek out more screen time of her own. Whatever message she takes from seeing Mom and Dad on screen all day, she is most likely going to increase her screen time as well.

In her 1972 poem "Children Learn What They Live," Dr. Dorothy Law Nolte reminds us that when it comes to raising kids, children learn what their parents model for them. For example, she tells us, "If children live with honesty, they learn truthfulness." Were she to update the poem, she might add the line, "If children live with adults who are on screens all day, they learn to be on screens all day."

Perhaps the worst part of a parent overusing screen time is that the parent is simply not there for the child. Mom or Dad might be physically at home, yet emotionally unavailable. Dr. Prooday describes it this way: "Technology also disconnects us emotionally from our children and our families. Parental emotional availability is the main nutrient for [the] child's brain. Unfortunately, we are gradually depriving our children of that nutrient." In chapter 1 we saw some of the ways in which Brett was replacing family time with screen time: he wouldn't talk to his mom on the way to school because he was on his phone. He attempted to exchange texts with his friends during dinner. He holed up in his room for the entire evening. When electronic gadgets become pervasive in all aspects of Brett's family life, it's not long before the effects show up in Brett's schoolwork.

## Helicopter Texting

Pushed to the edges of relevance by a child's obsession with screens, a parent can become desperate. Something teachers see in school far too often is parents who text their children *constantly* during the school day. The texts are very often things that can wait: "What would you like for dinner?" "How did your math test go?" "Don't forget to mow the lawn later," "Do you like the tuna in oil or in water?" and so on. Every other teacher I interviewed about this reported similar findings. I know about the content of these texts because kids will sometimes read them aloud when their phones go off in class. It often ends up providing a good laugh for everyone, which in turn discourages kids from turning their phones off. The sad reality is that because kids and adults are on screen so much of the time, these text exchanges might be the most a child communicates with his or her parents on a given day.

Texts about mundane matters are disruptive in class but might not be destructive on their own. In the long run, however, such constant contact between parents and their child during the school day can become quite damaging. The child learns that he cannot handle anything on his own. Normal conflicts that arise at school can leave a child feeling like his parent has to be there to fix the situation immediately. The student never learns to cope. Here is a common scenario: Something happens at school that has upset a student. Perhaps another student said that his shirt was lame. Perhaps a teacher scolded him for talking out of turn during class. Perhaps he forgot to bring his homework, and the teacher will not accept it late. Maybe an administrator caught him in a "hall sweep" for tardy students and he now has detention. Maybe a class discussion hit too close to home and brought up a bad memory. Whatever it was, his teenage righteous indignation has flared up. Feeling alone, he sends an angry, wounded, and sad text to his mom or dad.

Fifteen years ago, the trajectory of this story would have gone something like this: The kid processes the upsetting event for the rest of the school day, goes home, and talks it through with his parents. Or maybe he stews about it off and on at school and at home but eventually forgets about it, working it through on his own. Today, though? All too often, it goes like this: Before processing what happened, deciding if it was a serious slight, considering all the angles, or simply going on with the rest of the school day, the child immediately sends Mom or Dad a wounded text from school. Mom or Dad receives the text and perceives it as a distress signal. *Someone is hurting my baby!* This is natural. Parents are supposed to be able to decipher between "red alert" complaints from their kids and "blowing off steam" complaints. But most parents (certainly me included) struggle with that and often err on the side of a scorched-earth response. *No one is going to hurt my baby!* Angry phone calls or e-mails from the parent ensue, and now there is a struggle between the family and the school—and the child is caught in the middle.

Here's an alternate ending taking today's reality into account: The child sends the wounded text. The parent takes a deep breath and waits a beat. He or she sends back some variation of the following: "Wow, that sucks. I'm sorry that happened. Let's talk about it at home tonight. I love you." When the parent brings it up later, the child might say, "What text?" having calmed down and gotten over the slight. If, however, he is still upset, the parents and child decide as a family what to do: let it go, work on strategies for the student to handle it, or the parents might truly need to get personally involved. If the last option is necessary, at least it will be after everyone has taken stock of how important the event actually was. It will be with cooler heads. If it was an important event, waiting until the next day—in most cases—won't change the outcome.

When parents are in constant contact with their kids and handle every minor conflict as if it were World War III, they do a disservice to their children. They lose the ability to decipher major and minor events, and—perhaps worse—they never learn how to handle even little things on their own. That is not something schools or parents should have any part in teaching. If the child does not learn how to handle conflicts during the school years, what will happen later in life? What if something unpleasant happens when she is at college the following year? What if, as an employee, her boss says her performance wasn't good enough for a big raise? At what point does the student just have to process things on her own first?

Aside from the emotional and social stunting that can result from this constant parent-child contact, it can damage relationships at school as well. If a teacher has a student who exchanges multiple texts with a parent during class every day, it may well affect the teacher's opinion of that student and that parent. If every negative thought or experience a kid has is followed by a ranting phone call from an angry parent, then an adversarial relationship is bound to develop between that school and that family. That will not help the child.

## Technology Can't Replace Parenting

So am I saying to just let kids get bullied and pushed around by teachers, administrators, and other students? Of course not. However, I do have some suggestions for more successfully navigating the school-family dynamic in the digital age.

First, and perhaps most obviously, parents need to monitor their own screen usage, especially as it relates to their children and their schools. Are you "helicopter texting"? Do you really *need* to e-mail the teacher about a grade on the Civil War essay? This issue gets tougher when screen time is a major part of a family's life. Do your kids *expect* you to be in constant contact? Do they expect your responses to be immediate, even during the work/ school day?

Second is the need to monitor kids' screen time, especially during the school day. While this sounds obvious, it can be extremely difficult, depending on one's socioeconomic situation, among other factors.

Recently, I was on my way to the Outer Banks of North Carolina with my family. We had reached the point in the trip where we were all out of our minds, so we did what any red-blooded American family would do: we stopped at McDonald's. As we sat down with our food, I noticed a young girl sitting at a nearby table by herself, staring at an iPad. As we sat down, she immediately put the iPad down and with all the excitement of a puppy she came darting over to us, introduced herself, complimented my daughter's curly hair, asked why we were in town and what we did for a living, and told us about her favorite animals. When we asked about her, she explained her situation. Every day during the summer, she would accompany her mom to her work at this McDonald's. As her mom worked, she sat out in the restaurant alone with her iPad. She didn't mind, she said, because there was so much to do online. Moments into our conversation, her mom swooped in to "rescue" us from her extroverted daughter. We assured her we didn't mind and that her daughter was a delight. But as the mom returned to work and the daughter to her iPad, an older man in the booth behind us interjected his own assessment. "It's really sad," he said. "That iPad is her only babysitter."

Now that was during a summer day, not a school day, but the principle is the same: we have come to expect quite a lot from the screen in terms of child-rearing and communicating with our children. I don't pretend to know any more about the little girl in that story, her mother, or their financial situation. What I am imagining, though, is that the mom did not have enough money to pay for day care, or enough family around to watch the girl while she worked. In this situation, clearly the mother is doing the best she can, and telling her to do a better job monitoring her child's online time would be extremely insensitive.

However, there are two takeaways from this story. First, screen time is an issue for all of us, regardless of socioeconomic status, and it's something we all have to address in our parenting decisions. Second, screen time is an extremely convenient and inexpensive babysitter.

Here's another example of how we have outsourced to screens some opportunities to genuinely connect with our children. Long car trips used to be a time that families filled with talk, songs, and games. Of course, those trips were also filled stinky shoes, car sickness, and kids yelling, "Will you stop touching me!" and "Mom, he's on my side again!" It was not and is not all peaches and cream. However, these events become part of family lore. The discussions while cooped up together—even the tedious spats—become a family's cultural touchstones. Ask most adults today about a family vacation from their youth, and you'll not only hear about the Grand Canyon or Disney World, but you'll hear about the trip there. Those experiences are often important pieces of a family's puzzle. Will future generations talk about trips from their youth by saying, "I played seven hours of video games, watched YouTube, and never talked to anyone else in the car"? Chevrolet and other car companies apparently think that would be great. They have tapped into memories of long car rides and now advertise Wi-Fi in their cars. These ads show happy families cruising down the road. Of course, everyone in the car is isolated, but they all look happy. Of course there are times when it would be easier to use screens, such as on long car trips or in the grocery store. However, these are also opportunities for interaction. Perhaps a reasonable compromise is to set a screen time limit (rather than a ban) on a car trip, or to allow iPad use only during checkout in the grocery store. Limiting screen time is an uphill battle, for certain, especially if all the "cool" parents in your neighborhood allow unrestricted screen access. It's a matter of making it a priority, and that's tough because in modern life *everything* is supposed to be a priority. However, it will not only help your kids have healthier social skills; it will help them with all the issues outlined in this book.

## But What Should I Do Now?

So you've put your foot down about screen use, and something like this happens: While on that same trip to the Outer Banks last summer, we were staying at a beautiful resort. While checking in, I overheard the following conversation between a father and son as they walked through the lobby. The son appeared to be about ten years old.

Dad: "So I don't want you using this time at the beach playing *Pokémon Go*. I want you to take advantage of the opportunities to build sand castles and swim in the ocean and find seashells, because those things are fun. Right?"

Son: [silence]

Dad: "Right?"

Son: "No."

Many kids today simply are not sure what to do when they're not on a screen. They really don't know what to do outside. It might be tough to fathom, but in many cases parents need to teach their kids what to do outside and away from screens. The only options for a child cannot be screen time or else supervised practices or lessons. There has to be time for kids to simply be kids and goof around. We should be doing whatever we can to foster these non-screen goofing-around times. This, too, has benefits in school. If kids know how to handle themselves while not on a screen and while interacting with other kids, they will have more fulfilling relationships and will be more willing to participate in class.

### TAKEAWAYS

- Monitor and set sensible limits on screen time. If the average teenager is now spending nine hours per day on a screen, cutting that time in half might be a good goal. Parental monitoring of screen time has many benefits. In a study published in the *Journal of the American Medical Association Pediatrics*, Dr. Douglas Gentile explains that parental monitoring of screen time has ripple effects. The two main benefits are that it lowers total screen time and it lowers the exposure to violent imagery. The former has the additional benefits of improving school performance and increasing weekly sleep for kids. Reducing screen time and increasing sleep has an additional benefit of lowering the risk for obesity. Lowering the exposure to violent imagery has additional benefits as well. "Prosocial behaviors" increase and aggressive/violent behaviors decrease.

• Provide opportunities for kids to be kids. Encourage kids to spend time in pairs and groups with no screens and no adults. I am not suggesting leaving the house after tossing the key to the liquor cabinet to a group of teenagers who are downstairs with the lights down and the music up. I am, however, suggesting that kids need time alone to be bored and get creative. This is going to look different at different ages. Soccer practice is great. Piano lessons are wonderful. But there has to be time for kids to play on their own (when they're younger), or just hang out (when they're older). This has everything to do with their success in school. Kids who can figure out what to do without an adult or a screen directly leading them to the answer are much better problem-solvers. That does not often happen on its own. It has to be cultivated. The way to begin that cultivation is to allow for unscripted screen-free time with friends.

• Keep devices in common areas of the home. Kids are far less likely to watch and do risky things online if they are sitting in the family room with siblings and parents than they are if alone in their rooms. It's the same in the classroom. Proximity works wonders on student behavior. When students in a certain part of the room are having a particularly chatty day, most veteran teachers have learned to teach standing very near those students.

Often, the chatting doesn't even have to be directly addressed. The students naturally get quiet when the teacher is close. If our kids use their screens with us close by, our mere presence will serve as a reminder about what is acceptable.

- Monitor your own screen time in the house and model appropriate screen-usage behavior. This can be a tough one. Writing this book, for example, has provided many opportunities for my children to wonder where I am and what I am doing. We all have things that have to get done. We have bosses who put last-minute work on our plates. We have bills that have to get paid. We have e-mails that must be answered. What we have to focus on, though, is whether or not those are things that have to get done *now*. Not every e-mail or text needs (or even warrants) an immediate reply. Our friend will live if we don't immediately "like" his Facebook post about his new driveway. Monitoring our own screen time is one of the best ways to support our kids' healthy use of screen time. Psychologist Michael Oberschneider has written a children's book called *Ollie Outside*. It's about a boy who wants to go outside and play but everyone else in his family is consumed by their screens and won't go outside with him. The fact that this book exists is truly a sign of the times. The point of the book, though, is well taken. Being selective about our own screen usage will have benefits for kids both at home and at school.
- Talk with the kids. *Really* talk with them. It's much easier to eat dinner in front of the television than talking with the family at the table. "Talk time," however, is one of the key casualties of screen time. In an interview with CNN,

Dr. Ari Brown of the American Academy of Pediatrics said that screen time "reduces talk time between parent and child by 85 percent when the screen is on." This is particularly damaging for young children, who should be rapidly acquiring language and social skills. Most of us struggle with talk time, though, because it's more difficult. After a long day at school and work, many of us look for an escape. It's far easier to allow kids to play with their tablets, laptops, and phones while we're making dinner than it is to actively seek them out and attempt to engage them in conversation. But that's what they need. One of the saddest things I see in school is kids who are craving adult attention because they do not get it at home. Many kids just hang out after school. They go from teacher's room to teacher's room, often just looking to chat. The conversation will typically start with something related to class, but it often becomes clear that they are just looking to talk to an adult. I've seen a marked increase in this behavior over the last few years. Kids isolate themselves, their parents are isolated at home, and the need for attention has to come out somewhere.

• Stop texting kids when they are in school (part I). If you're trying to support your children, this one sounds counterintuitive. Most parents truly believe they are being helpful by texting their kids during the school day. They believe that staying in constant contact is the best way to advocate for their children. Parents expect return texts so they can immediately pounce on any potential problems. That's a problem for their kids. When parents are in constant contact, they do not allow their children to truly be present in their natural environment, which from 8:00 AM until 3:00 PM is school. Kids have to have a chance to formulate who they are. That is hard work, and some of it has to get done away from home. If they feel like they are never truly on their own, then they will never truly be ready to be on their own. Conflicts arise during any school day. Most of these conflicts are relatively minor. Solving those manageable difficulties is a critical part of growing up, and parents have to allow it to go on.

- Stop texting kids when they are in school (part II). The other problem associated with parents' constant contact during the school day is that the child is not truly ready to learn if she is in a daylong conversation with Mom or Dad. Before you hit Send on a text, ask yourself if it is truly a pressing matter. If you answer that question honestly, most of the texts during the school day will stop, and students will be freed up to be students. In my ideal world as a teacher, there would be no texts at all between parents and students—even in times of emergency. For generations, parents have called the school office and messages have been delivered to the student. That is still the best way to handle tough times and emergencies. That way, a trained counselor can deliver the news, or can bring the child to a private office to call home. Texting a child about a death in the family (I have actually seen this happen), for instance, puts too much on a kid in a far too public forum.

These suggestions are meant to support a child's most important relationships—those with his or her family. Relationships with others, and social skills in general, are the focus of our next chapter.

## 7

# Revitalizing Social Interaction

*Our networked life allows us to hide from each other,*
*even as we are tethered to each other.*
—Dr. Sherry Turkle, professor of the social studies of science
and technology at Massachusetts Institute of Technology

ONE DAY I WAS HAVING my students play a review game in preparation for a test. They needed to be arranged in teams of four and five students each. When I allow students to pick their own teams, as on this day, unsurprisingly they gravitate toward their friends or classmates with whom they've worked in the past. On this particular day, however, one girl was having trouble connecting with any partners. Some students, including her friends, were on a field trip, leaving her without her usual group. She stood up, frantically scanning the classroom for someone she knew. I watched her confront the issue. She was frozen.

I was perplexed. This is a student who regularly contributed to class discussions, obviously had plenty of friends, and seemed very well adjusted. She was curious, funny, and appeared to be self-assured. There was a team on the other side of the room in need of one additional member. They let that be known, but the girl stood in place. A student on the team called out again, "We need one more if anyone doesn't have a team." The girl remained still. I finally said to her, "Why don't you join in over here? They need you." She looked horrified, but swallowed hard, nodded, and walked toward her new team with all the enthusiasm of a prisoner approaching the electric chair. It was what she did on the way, though, that really caught my attention. As she started walking, she reached into her pocket and pulled out her phone. She didn't check for messages. She didn't even look at it. She just clutched it. She eventually finished the journey across the room and sat down in the empty chair. The other students greeted her, she quietly returned their greeting, and her knuckles became slightly less white as she loosened her death grip on the phone. After a few minutes, she put it back in her pocket and was once again her animated, personable self. The game went on. I've forgotten how her team did.

## My Life Is My Phone

The profundity of this student's action hit me. Feeling a little anxious, instead of drawing on inner resources this young lady turned to her phone. In that moment, it was her security blanket. She just needed to know it was there. What must have been going on in her autonomic nervous system to make her react that way? Every teacher I asked about this said it sounded familiar. What I find troubling about this is that what the gadget represents replaces even the information on the screen itself. Students are increasingly unable to navigate their world and their relationships without their devices. Many of our kids today echo the sentiments of a participant in a 2013 survey of the effects of their phones on the mental health of college students: "My life is my phone," the student said. If students truly feel this way about their phones, and an increasing number of studies suggest that they do, how does that affect their relationships with one another and their families? More important, how does the excessive use of their phones and screens in general—in and out of school—affect not just these relationships but all of their social skills?

## Alone Together at School

In my school, every other day, students have forty-five minutes of mostly unstructured time. This is a period that students may use as they wish, provided they are not getting a D or an F in any class. Students use this time to do homework, make up tests and quizzes, get extra help, or just hang out with friends. The school has been providing this time to students for about a decade, and it has been successful by almost any measure. I typically had a few dozen students who would show up. For the first several years there were usually lively discussions of important world events (read: weekend plans). To be certain, students would often study and get extra help. Just as important, though, they would use that time to get to know one another, their teachers, and their school community.

Slowly, though, over the years the room got quieter and quieter until it was often utterly, eerily silent. Every student would have his or her devices out—phones, tablets, laptops, e-readers, and so on. Most would have earbuds in, and only rarely would they be talking with one another. One day, I told students that they were still welcome to come to my room for that period, but they were no longer allowed to use their devices. How many students now typically show up (who are not seeking extra help)? Zero. This is not OK. This is not an innocent change. The fabric of our school community is changing for the worse, and it is a direct result of students' use of technology. Again, this overreliance is at least in part driven by schools themselves.

In recent years, many schools have adopted a BYOD (bring your own device) policy, which officially lifts the age-old ban on personal electronic devices. These policies actually encourage students to bring their cell phones, laptops, iPads, and other screens to school. In fact, now many school systems go a step further and provide a device for each student. Many more schools will soon do the same, in the absence of efforts to prevent it. When we encourage the use of screen time in the classroom, we accelerate the deterioration of our school communities and the ability of our students to navigate social situations.

In large, transient school systems, it is common for students not to know their neighbors outside school. In school, therefore, they do not find it odd that they do not know the person sitting next to them in class. As a result, it becomes easier for students to justify focusing on their phones when they have a few free minutes. Rather than interacting with one another, students turn inward and focus on the screen they are carrying. However, this is an issue in smaller, more tightly knit communities as well. In 2009, for instance, the school board in Jewell County, Kansas (population around 3,000), banned cell phones from schools. Eric Burks, principal of Jewell High School, had worked as the county's technology instructor. Seeing what cell phone usage was doing to his school community led him to make the decision, which was supported by the local school board. A sense of community is valued by large and small school districts alike. That sense can only be fully developed in a school if students are regularly, personally engaged with one another.

Dr. Dania Lindenberg of the Scripps Coastal Medical Center says, "Children spend far too much time in front of a screen. This takes away from time spent interacting with their friends and family, being physically active, and reading or doing homework. Coordination, social skills, and problem solving are best developed away from screens." In spite of this, teachers are routinely encouraged to use *more* digital technology in class. Notice the first casualty of screen time that Dr. Lindenberg lists: interacting with friends and family. This includes teachers and peers at school. Navigating social situations is difficult enough for kids. What students need is more practice interacting with one another and their teachers. They do not need any more practice texting, playing video games, or watching YouTube. At these activities, they are already experts.

As is true with regard to the deleterious effects digital technologies can have on critical thinking, focus, and problem-solving, modern research leaves little doubt about the harm digital technologies are having on the ability of young people to develop socially. The implications for schools are profound. Psychological researcher Douglas Gentile and his colleagues found that children who are heavily involved gamers (those who play video games thirty-one hours or more per week) are significantly more likely to suffer from anxiety, depression, social phobias, and lower school performance than moderate gamers (those playing nineteen or fewer hours per week). The anxiety, depression, and social phobias suffered by these students will take a toll on their schoolwork. Thirty-one hours of video games per week seems extreme. However, it is not just children at the fringes who are suffering social harm from the overuse of screen time. In a 2010 University of Bristol study, Dr. Angie Page found that children who have more than one to two hours per day of screen time show a 60 percent increase in psychological disorders. Nearly all students today have far more than one to two hours per day. Further, in 2013 University of North Carolina at Chapel Hill psychology professor Dr. Barbara

Fredrickson and her research team found that a person's ability to develop friendships is biologically diminished the more he or she replaces face-to-face human interaction with screen interaction. They found that *at the cellular level* we are changed by our habits and actions—especially with regard to screen time. If we do not actively seek out connections with other people in real life, we actually lose our ability to make them.

So what does this research have to do with schools and education? If education is about students mastering content and skills, what does it matter if they do not feel connected to their school, their teachers, or one another—as long as they are learning the material?

For starters, Lieselotte Ahnert and her team at the University of Dresden and the University of Vienna found that "cognitive processing is much more effective if close teacher-child relationships are involved." Further, as referenced earlier, Hunter Gehlbach's team at Harvard found that when students find similarities between themselves and their teachers, they perform better in class. It shouldn't be particularly surprising that students do better in classes where they like the teacher and the teacher likes them. Most rational people would rather be around and work with people they like, as opposed to the alternative. That's human nature. What's interesting about these studies, though, is the finding that students' brains actually work better when they perceive close relationships or similarities with their teachers. It's as if, in classes where they like the teacher, students could not do poorly even if they wanted to. That sounds like pretty good stuff. The obvious conclusion is something good teachers have known for a long time: teaching and learning are much easier and more successful when there is strong teacher-student rapport in the classroom. Schools need to do whatever possible to encourage the social skills of relationship, rapport, and community building.

As we have seen, student overuse of screen time is leading to a deterioration of these critical social skills. In fact, University of California, Los Angeles, psychology professor Dr. Patricia Greenfield has found that heavy use of screens causes young people to lose the ability to understand the emotions of other people. The good news is that she has found a fix for this. Several days unplugged and away from screens can cause a dramatic increase in the ability of kids to recognize the emotions of others and become more empathetic. So what are we doing in schools? Putting them in front of screens all day, of course. Who wants kids to be able to recognize all those pesky emotions, anyway?

Teachers see the effects of this every day. A teacher I interviewed taught psychology at a suburban high school. She was a twelve-year veteran and, by all accounts, beloved by students and staff alike. During one unit each year, she had her students engage in a group activity in which they were to imagine there had been some sort of apocalyptic event (a world war, meteor impact, computers taking over education, etc.). The teams were told they were going to have to live for a while in an emergency bunker. They were then given a list of ten people with various personalities and told they had to reach a consensus on which ones they would allow into the bunker. They could only admit five. The point of the activity was not which ones they would admit, or even what their value system was. It was to discuss group dynamics. The students were the subjects of this simulation, though they didn't know that until it was over. She used this activity to bring out the various roles that people play in groups: leaders, facilitators, appeasers, inhibitors, and so on. Invariably, students would get extremely animated in their discussions of which people to include in the group. There would be good-natured banter as the teacher led students in a discussion about their own roles. Students would recount what happened in their groups, and they would all identify what roles people were playing. The lesson went on for a full ninety-minute class, and could have gone longer. It was engaging and valuable, and the teacher looked forward to it each year. Students learned the concepts well, and it was one of their favorite lessons.

Beginning a few years ago, though, this activity changed. The group discussions about whom to include got shorter and shorter until, the last time she did it, the longest any group took to reach consensus was four minutes. The class discussions were even shorter. One might hear that and say, "Oh, isn't that great—students these days know how to reach consensus quickly." They reached consensus, for certain. However, for those last few years as they engaged in the lesson the teacher said that, as she walked around the room to the various groups, she would hear students saying things like, "I don't really care. Just pick something so we can be done," or, "This is stupid." This teacher, as many do, had a policy that permitted cell phone use when work was completed. So once students were done, they knew they could take out their phones and resume watching movies and amusing cat videos on YouTube and sending very important selfies to one another through Snapchat. In a final attempt to save the lesson, the teacher abandoned the policy and said phones were no longer allowed at any time. This, however, did not have the desired effect. Students would still say, "Just put something down so we can be done." Then they would sit in silence, despite the teacher's best attempts to get students to think about the personality types and implications of the question at hand. During the class discussion, too, students would be mostly silent. Finally, unwittingly, one of her students put his finger on it. He said, "We don't know what roles we were playing because we weren't playing. You didn't tell us who was in charge and we didn't know which personalities you wanted us to save, so we didn't really do it."

Now this hits another issue—the fact that, in the age of constant "high stakes" testing, students are increasingly incapable of thinking on their own. They just want to know what is on the test. However, for our purposes, the social implications of this lesson are tragic. Students could not have an organic discussion about something without being told how to do it. Does that mean that screen time ruined their ability to interact socially? Not entirely. However, it clearly has not helped. Students spend more time than ever interacting with their screens and are losing the ability to converse in any sort of genuine, face-to-face way. As a sad postscript, that dynamic course is now dead. Word got out that it involved a lot of talking and discussion and, what's worse, no cell phones were allowed. As it was an elective, interest dried up. This teacher went from having five full sections with a waiting list to get in to having the course die out entirely. The tragedy is that a course in psychology is exactly what digital natives need to understand the importance that social interaction has on their own well-being.

Having open, honest dialogue is difficult. It's difficult for well-adjusted adults and it has always been particularly difficult for kids of all ages. However, I can say without question their ability to have open, honest, face-to-face, socially appropriate dialogue has deteriorated over the last few years. There is no doubt that the rise in screen time is at least partly to blame. I would argue it's mostly to blame. There are many excellent books out there on the subject: *Reclaiming Conversation: The Power of Talk in a Digital Age* and *Alone Together* by Sherry Turkle and *The Mobile Connection: The Cell Phone's Impact on Society* by Rich Ling are a few great places to start.

What we're more interested in here, of course, is how deteriorating social skills affect a child's education. Students unable to function in relationships and have genuine conversations are not only going to be socially deficient. They are going to be behind in school and in the workplace. Think about your own job. How successful could you be if you could not actually communicate with other people? There are likely jobs where that is possible. I just don't know of them. For most of us, communication is critical.

## Who Needs Teamwork When You Can Google?

John Singh, a managing director at a major New York investment bank headquartered on Wall Street, told us that communication skills are critical in virtually every aspect of what they do. In fact, he says, they were critical when he was an undergrad at the New York University Stern School of Business. In every class he can recall, he was required to work as a part of a team on projects and presentations. The rationale from professors, he explained, was that students needed to get used to it. While at Stern, he was told early and often that communication was the key to success—both at school and in the business world. Successful he was, graduating with a business degree in 2004 and eventually with an MBA from the Wharton School of the University of Pennsylvania. At Wharton, all his classes during his first year involved being part of a learning team. Since he has worked on Wall Street, he says the social lessons he learned in school—particularly how to navigate complex group dynamics—have been critical to his success, as well as the success of virtually every one of his colleagues. In regard to his work, he says, "Everything [we] do requires team work, collaboration, and communication." Perhaps you're thinking, *But my kid isn't going to go to business school or work on Wall Street.* Perhaps not. However, in nearly all jobs today "soft skills" such as influencing, negotiating, and motivating are absolutely essential. But these skills are not just about being successful at the next level. What about K–12 education?

I'm glad you asked. I heard this story from a colleague in a different state who teaches AP economics. In that class, he does a classic simulation in which students are divided into teams. The winners get candy, so students behave pretty much as if their lives are at stake. Students are divided into teams of various sizes and their task is to cut out a predetermined shape from construction paper. They have to cut out as many as they can in three minutes. Each team has two pairs of scissors regardless of the size of the team. The winning team has the highest labor productivity, which means output per worker. When the simulation starts, students furiously fold, trace, and cut. After three minutes, each team counts the number of shapes it was able to cut out and divides by the number of team members. Even if the team of two has the smallest number of cut-out shapes, it only divides its score by two and so invariably has the highest productivity. This activity provides a great introduction to the concept of labor productivity and how important it is for an economy.

This teacher had done this simulation over one hundred times. The team of two had won every time—until this past year. This year he started the simulation and explained it as he had every other time. It started off just fine. However, in one class, the team of two was having trouble. The teacher noticed that after about thirty seconds they were just looking at their phones, not cutting anything at all. Because the point of the simulation is shot if the team of two doesn't win, he went over to investigate. First, he had them put their phones away. Then he asked what was going on. "Well," one student explained, "we divided the tasks. I was going to fold the paper and trace the shapes and she was going to cut." The teacher told her that sounded like a good plan—division of labor and all. She continued, "But she's left handed and these are right handed scissors." Ahem. "Are you right handed?" he asked. She said she was. "Do you see a solution here?" he asked. The left-handed student chimed in, "Ohhhhhh . . . maybe I should fold and trace and she should cut!" They quickly changed tasks, but too much time had gone by. For the first time in the long history of this simulation, the team of two lost.

There is so much wrong with this I hardly know where to begin. It is obviously an indictment of their problem-solving abilities. However, these were not students who were struggling. First of all, they were taking AP economics, which is an elective. This is the sort of class that tends to attract highly successful students. Who in their right mind would take such a class if they were struggling? In addition, these two were performing better than the class average at that point, and continued to do so. Their real failure was in their social interaction. The two students knew one another well. Their friendship went back to long before the start of the class. They always sat together. It was not simply that they didn't know one another and therefore didn't talk through the problem. It was that, even though they knew one another, they didn't talk through the problem. The left-handed student simply said, "Oh, these are right-handed scissors. I can't use them." The other student said, "That sucks." Then they both said, "Oh well," and got their phones out. One student said she Googled "how to win a productivity simulation" on her phone. When that did not immediately yield a satisfactory answer, she went to her default: Snapchat and amusing videos. The point is, students have to know how to work together to solve problems in school, at work, and in their personal lives, and they are losing the ability to do so.

## Why Texting and Snapping Aren't Talk, and Why It Matters

Not being able to work together and navigate social situations has consequences that reach beyond the classroom. When students are not able to have face-to-face conversations, they resort to texts, Tweets, and Snapchat. However, these digital forms of communication are one dimensional when compared to face-to-face interactions. They lack all the nonverbal cues that augment the message being conveyed, such as tone and body language. These digital forms of communication also allow the person delivering the message to avoid having to experience the recipient's reaction. But this experience is key to developing social etiquette. Teenagers are awkward for a reason. They're experimenting with different ways of interacting with others. Without the instant feedback they receive from face-to-face interactions, they never learn what's acceptable and what isn't, and consequently they never emerge from their awkward stage. With their digital interactions, they don't have to see the person's hurt when they text them "You're ugly" or their anger when they post "You're a slut." That human empathy that kicks in to prevent mean kids from growing up to be mean adults never fully develops.

Yet this is how kids "talk" to one another. They actually use the word *talk* to describe the text exchanges they have with their friends. I routinely hear my students and my own children say that they are "talking" to one another, when in fact they're on their phones texting back and forth. Of course, texting is very convenient and has a place in this world. When it becomes the primary mode of conversing, however, it does not allow for actual conversation skills to develop—skills like using and deciphering tone, facial expression, body language, vocal cues, and more.

I wrote in chapter 1 that Brett had sent eighty-seven Snaps during the school day. That likely puts him on the low end of his peer group. In the last year or two I have noticed that my students spend an inordinate amount of time in classes, in the hallway, and in the cafeteria looking at their phones and making faces. Because I'm old I at first had no idea what they were doing. One day last year I was walking around the room helping students with a problem set. A few students had completed their work and one had her phone out. She was making faces at it. I walked behind her seat to see what she was looking at: herself. She was staring at herself. But it was more than idle staring. She was taking selfie after selfie. When she took one she liked—and the dozen or so I saw her take were identical—she sent it to another student. The other student was in the same class. I know this because within five seconds a selfie from this other student popped up. I watched them send a few more, virtually identical pictures to one another.

I went to the front of the room and stopped class. I didn't name names, but I described what I had just seen. They looked at me with blank faces as if to say, "Yeah, and . . . ?" I asked students if this was normal, and how many students had used Snapchat to send a selfie during that class period. More than half of the hands went up. I asked how many had sent at least one selfie during that school day, while in any class. They laughed. Every hand went up. One student said, "I've sent two since you asked if this was normal." More laughter. I was laughing as well, but for a different reason. "This is really how you communicate?" I asked. They were the teachers now. They told me that it was "easier" to "talk" that way. I pointed out the absurdity of calling it talking. They pointed out that there wasn't yet a good alternative word for it. They said they liked texting and Tweeting and Snapping because you had time to respond, and in a face-to-face conversation you don't always know what to say. I asked if they felt like they needed to reply immediately when someone sent a message. They said it was rude not to reply right away. I questioned the logical inconsistency of saying that text was better because it gave them time to respond, but they felt pressure to reply immediately. Yet this seemed to make sense to them.

Part of this, to be certain, is simply a generational thing. They do this; adults don't. The end. However, there is no denying the ill effects of communicating in this way. Think about what that would do to you—as an already self-conscious teen or tween—if you took literally hundreds of pictures of yourself per day (which many tweens and teens do). "Narcissistic" doesn't even begin to scratch the surface. What reasonable person could possibly want to look at his or her own face that much? The answer is that no reasonable person would. They're kids and we love them, but "reasonable" is not always a word that leaps to mind when thinking of how to describe them. School-aged kids are working on being rational and reasonable, but they're not fully there yet. So imagine that world. Everyone you know is bombarding you with pictures of their faces all day long—before school, in school, after school, during dinner, while you're doing homework, at bed time . . . all day. If you *didn't* respond with pictures of your own face, you would likely feel like the weird one. Teens and tweens especially don't ever want to feel like the weird ones, so they practice their selfies and send them out. Once you're sending hundreds of pictures of yourself, imagine the message your brain gets: *I am so important and people love my face so much that I must send out a hundred or more pictures of it every day just to give the people what they want.* That's not socially appropriate or mentally healthy.

The crazy part is that Snapchat is just the beginning. "Social media" has already been the subject of many books, movies, and television shows. While I cannot give it a full exposition here, it certainly bears mention with regard to education. For school-aged kids, Snapchat is simply the flavor of the month. When it first came out, it was Facebook (or Myspace before that). That is now passé for most kids. Instagram was hot for a while. Vine was really important to kids for about twenty minutes. As of now, Twitter and Snapchat are still big. Chances are good that by the time this book comes out more social media apps will have risen and fallen. The point is not the particular app. It is social media in general, and most students are heavy users. As you probably realize, there is very little that is truly "social" about social media. These are applications with which one interacts while alone. That is true for anyone of any age. However, if you're an egocentric teen or tween, that can be dangerous.

Imagine being said egocentric fourteen-year-old with an electronic device. You go out to dinner with your family. You order a steak. When it gets to your table, it looks amazing. It looks so amazing that you just *know* everyone will want to see it. So you take a picture of it and post it to the social media app of the moment. During dinner, you check to see the number of "likes" or "+1s" your picture gets. As the bill is paid, you are disappointed to find that only two people have liked this picture. You have dozens of connections through this app. *Why were there only two likes? What is wrong? Why don't people like me!?* That's what goes through their heads. It was a picture of a piece of meat, and students derive information about their self-worth based on whether or not other people clicked a thumbs-up or +1 icon when they looked at it. If you're that child, though, you see it as a slight. You feel embarrassed and vaguely angry. You have several friends you know should have liked that picture, and they didn't. You'll spend the next day at school avoiding them. When you get to history class and are supposed to do a reading about Aśoka, you cannot possibly pay attention. Only one of the thirty-two pictures you've posted to Instagram today has gotten double-digit likes. One of your friends with whom you had a forty-seven-day Snapchat streak has let it die. Your social capital is in the toilet . . . You cannot be bothered with Aśoka, or any other Mauryan for that matter.

This is what teachers are dealing with in school every day. We know kids are egocentric. That's developmentally normal. However, their struggles to be understood and well liked are compounded by incessant bombardment with and involvement in social media.

Many students are even more consumed with what others say about their online personae than they are about what people say about them "IRL" (that's "in real life" for you digital immigrants). Research bears this out. Tagrid Leménager and her team at the Central Institute of Mental Health in Mannheim found that heavy screen users actually *prefer* to get positive feedback about their online avatars than to get face-to-face, live compliments.

With schools pushing more screen time on kids during the day, it's just a hop, skip, and a jump to get to the fact that schools are helping to create a generation of kids who would rather get an online like than a compliment from someone in person. If a person gets more reward from an online comment about his or her online personality than from real-world positive feedback, it only makes sense for that person to spend more and more time online and care less and less about interacting with live human beings. The movie *Her*, in which Joaquin Phoenix's character falls in love with his computer operating system, no longer seems like science fiction.

## Stop Fooling Yourself—Your Kid Looks at Porn Online

Speaking of love and relationships and how screen time affects a young person's perception thereof, it's time to address the naked elephant in the room: porn. The social effects of pornography on young people are well documented, and they are devastating. It is wishful thinking of the most dangerous kind to provide teens and tweens digital devices and expect them to ignore the sewer of the Internet. Do you remember being twelve? Or fifteen? Or seventeen? Now imagine if the adults in your life said to you, "Now, Jordan, we're going to give you a very powerful device. This device can access all the most engaging entertainment you can possibly fathom—video games, television shows, movies . . . and more pornography than you could possibly view in ten lifetimes.

Make sure, though, that you don't look at any of that. Please just look at your geometry teacher's screencast on the Pythagorean theorem, and then watch your Latin teacher explain deponent verbs." What would you have done? Me too.

This is an issue at home, at school, and at work. This is obviously a potential issue if your child has access to a device and the Internet at home. However, we compound the problem at school when we put them in front of screens there as well. One would think that schools would block inappropriate sites, and they do. However, if even one motivated kid figures out how to get around the school's firewall, then every kid will know how to get around the school's firewall—and they do. This is true from elementary through high school. The average teenage boy watches fifty pornographic video clips per week. That's primarily at home, but it happens in schools every day. Kids come to school with all sorts of things loaded on their phones, pornography included. Even if it's not loaded on a phone, it's certainly available if a student uses his or her own data plan rather than the school's network. In summary, short of banning all devices, there is virtually no way to keep our kids from accessing online pornography at school.

Research shows that males are five times more likely to view pornography than females. Girls, though, are not immune to the damaging effects of more and earlier sexual messages and images. They are far more likely to be victimized by hypersexualized advertising, for example, which research has shown leads to risky behaviors, depression, and anxiety. *American Girls* by Nancy Jo Sales addresses issues surrounding the damage that social media and technology is doing to girls. Still, you might be thinking, *So what? Kids, boys especially, have been fascinated with pornography as long as people have been having sex.* The difference, of course, is in the availability. A few decades ago, if your dad didn't subscribe to *Playboy*, then you had to wait for the JCPenny catalog to come to your house. (I mean, that's what I've heard.) Today, though, the most lurid, depraved, and violent acts are a click away. Why should that concern us when it comes to our kids' relationships?

Consider a few facts. First, after analysis of hundreds of scenes of pornography from the most often viewed videos, it was determined that 88 percent of the scenes contained acts of physical aggression, and half contained verbal abuse. If we want young people to understand what a healthy relationship looks like— sexual or otherwise—we need to do whatever we can to keep them from seeing physical and psychological abuse as "normal." The more our young men (in particular) watch porn, the more normal these behaviors seem to them.

Second, according to the *Journal of the American Medical Association*, there are three troubling effects on the brains of regular porn viewers. First, there is a negative relationship between the amount of time spent watching pornography and the amount of gray matter in the brain's reward center (more porn equals less gray matter). Second, when a person's brain becomes used to watching a certain amount of pornography, she or he will have to watch an ever increasing amount to get the same reward response from the brain. Third, more pornography viewing leads to a weaker connection between that reward center and the prefrontal cortex, which governs decision making. So, in short, the more porn one watches, the less complex one's reward center becomes, the more porn one needs to watch, and the more likely one is to make poor decisions.

Third, the phenomenon is extremely widespread. Before the age of eighteen, 93 percent of boys and 62 percent of girls have had at least some exposure to online pornography. Again, the only realistic way to stop this is to limit the amount of screen time our young people have. We already know their impulse control is not fully developed. Simply expecting them to not look at pornography isn't going to work. Yet in schools we provide devices and time for them to be used, and we demand that they use them at home. It's not a great combination from a porn-reduction standpoint.

When it comes to romantic relationships, the effects are disastrous. Young people end up seeing physical and psychological abuse as normal parts of intimate relationships. They see the emotionless sexual acrobatics in many videos and assume that is how men and women typically make love. That is going to leave them feeling either inadequate, abnormal, or with a warped view of what should go on in the bedroom (if not all three). This last effect might be the worst, because distorted and even destructive sexuality is what the young people might emulate or come to expect from a partner.

So, that's depressing. The good news, of course, is there are many steps we can take to help our students maintain and repair their social functioning. These steps will not only help relationships for the sake of the relationships. Healthier social attitudes will help our kids be more successful in school. When kids develop distorted views of relationships, it makes working as a productive member of a team or group much more difficult. It makes connecting with teachers and adults more difficult. None of that will help students be the best versions of themselves.

### TAKEAWAYS

- Encourage kids to work with a partner or partners when they get that option. Many teachers will offer this choice to students on certain assignments. We all need to do a better job of giving kids chances to collaborate in person, not online with tools such as Google Docs.

This provides critical practice at honing social skills and working with a team to tackle a problem. We need to help kids choose to reach beyond themselves and work with others.

- Invite your child and his friends to work (and play) at your house. If nothing else, this will give you insights as to how your child and his peers interact socially. You have likely seen a group of kids sitting around staring at screens together. They might look like they're having fun, laughing at what they're finding and sharing that with one another, but such interaction is fundamentally different from communication when no screens are involved. Even if kids are loud and animated when using screens, they're still looking at the screen and not into the eyes of another live human being. They are not learning how to read body language or facial expressions. Of course, simply saying, "No screens for anyone who comes over," might mean that your child becomes the neighborhood pariah and you appear to be the crotchety old person. You don't want that, and neither does your child. However, a balance must be struck. Give your kids as many opportunities as possible to work and hang out with other kids, and watch them interact. That will give you the insight you need when setting sensible rules for your home.

- Forbid screens in kids' bedrooms. One rule that is used by *every tech executive* in a recent survey is that their kids are not allowed to have any screens in their bedrooms. Aside from the obvious benefits—being able to more closely monitor what a child is doing online—this rule has the benefit of keeping the child in the common areas of the home. That is, even if the kid is going to be playing a video game or watching a movie, she will have to do it with other people around. This will simply give more chances for social interaction—direct and indirect. It also reminds us as parents how long our kids have been on screen. If a child is on a screen in her room, not only do we not know what she's doing, we often forget how long she's been doing it. We are busy doing laundry, making dinner, or whatever—and we forget that our children are on screen that entire time. If they are sitting out in the open, it is a reminder for us as parents as well. Kids, even teenagers, like talking to their parents and siblings—not every second of the day, to be certain—but they do. Like so many good habits, learning how to communicate well and interact in socially appropriate ways starts at home.

- Replace screen time with friend time. I heard this idea from a colleague, who worried that her kids were choosing to isolate themselves and stay inside rather than play with friends. She told her kids that they could have as much screen time as they want—with one catch: for every hour they want on screen, they have to spend two hours with a friend or friends off screen. Genius, right? When she announced the plan to her kids they were excited. She put a chart on the refrigerator and each child kept track of his or her time with friends. For the first few days, her kids would hang out with friends for a few hours and then race gleefully back inside to soak up their hard-earned screen time. Because they didn't want to continue burning through the same friends, they were calling ones with whom they hadn't played in a while.

They were even finding phone numbers for acquaintances from school and asking to hang out. Within a week, they had more plans with friends than they could possibly fulfill, and screen time stopped being a big issue. An added benefit to this solution is not only do her children now have more friends, they also have more potential partners for work at school, and through play they've practiced social skills and problem-solving that are critical for their education. This suggestion might not work in every home or neighborhood. However, with some creativity it can be modified to fit many situations.

- Bring your children to school events. In most communities, there is no larger weekly social gathering of humanity than a high school football game. It's a great place to go if you want to see kids being kids, cheering on their team, and families bundled under blankets, snacking on popcorn and hot dogs. Of course, you can also see some not-so-great behavior: overcompetitive parents yelling expletives at referees and so on. That's just it, though: it's your community, warts and all, and kids need to be a part of it. If football isn't your thing, try attending a volleyball or field hockey game. Don't like sports? How about a theater production or an orchestra concert? The point is the schools in our communities offer many chances for families to be together. This is how our kids get to know where they live and who their neighbors are. Going to events like these sends the message that the school community is important—and it's not just high schools that host them, of course. Elementary schools have band and choral concerts and other special events. Middle schools have these, plus some sporting events. Your outings might inspire your child to get involved with some of these groups. Our children are more likely to be plugged into the community if they know it is important to their family. The common denominator of all these events is that they take place in the actual world, and involve no screen time. Helping our kids connect more deeply to the place where they spend a large chunk of their lives can, by definition, almost improve their social interaction skills.

- Talk with the kids (again) about their use of screen time. If your children are using social media to "connect" with friends, be aware of that. Employers routinely check on the social media presence of job applicants. Know what social media apps your kids use and be "friends" with them on any online applications they use so you know what your kids are up to online. In 2010 the Kaiser Family Foundation found that a substantial majority of children have *no rules* on what they can do online. The solution to this is hidden in plain sight. Simply talking with your kids about their screen time and the social effects it has provides two clear benefits: first, it gives kids practice at interpersonal communication. Second, it gives parents some insight into what their children believe about relationships and social connections and how those are affected by their screen time. Their answers provide an important jumping-off point for any corrective actions moving forward.

As we will see in the next chapter, these corrective actions can help children at school—regardless of where they are on the achievement spectrum.

# 8

# Technology Is Widening —Not Closing— the Achievement Gap

*They haven't used it. We limit how much technology our kids use at home.*
—STEVE JOBS ON HOW HIS KIDS LIKED THE IPAD

"I DON'T HAVE ANY CONTROL OVER MY SON," a tearful mom admitted to Gabe, one of her son's teachers. Gabe couldn't help feeling sorry for this mother, asking someone like him for advice about how to raise her child. What insight could he possibly give her? At twenty-four, he had been living on his own for only a few years and was barely an adult himself. He knew nothing about raising kids, only about teaching them. But he could tell this mother had reached her breaking point.

Gabe had been warned about the boy, Chet. The school counselor had approached him before the school year started, saying, "We really have to keep an eye on him." Chet was new to the school. He had spent much of his eighth-grade year in an alternative school after being expelled from his middle school for participating in gang initiations in the locker room and stealing his mom's car. Gabe's school is relatively affluent, with a mostly middle- to upper-middle-class student population. Dealing with students with gang ties and a carjacking arrest is pretty rare. Gabe worried that his classroom management skills weren't strong enough to handle a kid like Chet. After psyching himself up to handle this "monster," the moment finally arrived. He spotted Chet approaching his classroom. His heart raced as he began to mentally prepare potential responses to whatever terrible thing Chet would say to in an attempt to challenge Gabe's authority. "Good morning, sir," Chet said.

*Nice try, buddy*, Gabe thought to himself. *Your little Eddie Haskell bit isn't going to work on me.* But this "act" didn't let up. For weeks, Chet continued to be a sweet, polite young man. Although Chet seemed a little rough on the exterior and his grades were pretty low, Gabe was relieved that he wasn't openly defiant or disruptive in class. You could even say that Chet and Gabe got along quite well. This is probably why Gabe looked the other way when Chet used his cell phone in class. For one, he wasn't bothering anyone, and two, Gabe felt a confrontation would sour their delicate relationship. But as the weeks went on, Chet was spending more and more class time playing games and not paying attention. By the end of the first quarter, Chet had an F in the class. The two had several friendly discussions about how Chet's lack of attention was contributing to his failing grade. But it was having little effect.

His mom complained that when he wasn't in school, Chet was online constantly at home. Of course, he was spending almost no time doing anything productive, like studying or homework. She was a single mom who typically didn't get home from work until several hours after Chet got out of school. By the time she walked in the door, he was firmly entrenched in his games and she was too exhausted to fight him. And when she did try to limit access to his devices, he would typically lash out at her. Like the young girl in the McDonald's, his electronic devices became his babysitters. She rationalized this by telling herself, *At least he's not out doing drugs or having sex.* His games were keeping him out of trouble, or so it seemed.

It became clear to Gabe that Chet's mom and he had similarly low expectations for Chet, with their only goal being for him to stay out of trouble. Allowing him to play on his technology helped Gabe and Chet's mother to achieve this goal. But what about his academic performance? By the end of the first semester, it became virtually impossible for his grades to go any lower. In an attempt to pull him out of his downward spiral, Gabe stopped handling Chet with kid gloves and tried to take a hard line on his phone use in class. At first Gabe would ask him to put it away, and when he wouldn't, Gabe would take it from him. Chet quickly started to resent Gabe for it, and their good rapport went out the window. Chet started showing up late for every class and only acknowledged Gabe to say something disrespectful. It became clear to Gabe that he had waited too long to do something.

## Unable to Get Ahead

Chet seemed to have the cards stacked against him. He belonged to all the groups that historically struggle in school. He was raised by a single parent, was in a lower socioeconomic class, and was an ethnic minority. More recently, you could add "male" to this list of struggling demographics. The disparity between these groups and their whiter, wealthier counterparts is known in the academic world as the "achievement gap."

The achievement gap has become the bane of school systems all over the United States. Closing it has been the holy grail of academic reform policy since *Brown v. Board of Education*. Initially, in the decade following desegregation, there was a significant closing of the gap. But in the 1990s it stopped closing. Despite the many efforts made over decades of policy reforms, now the gap is beginning to widen again. Lower socioeconomic students and historically oppressed minorities are doing worse in almost every conceivable measure of academic success: performance on standardized tests, GPAs, high school graduation rates, college graduation rates, college graduation rates among students with similar GPAs and scores on standardized testing, and even participation in extracurricular activities. The trend of income disparity has become significantly more pronounced in the last twenty years. In Chet's case, his issues in school obviously went deeper than his technology use. However, the problems resulting from these issues were clearly exacerbated by it.

For example, because his mom worked late hours to provide for her family, Chet was left unsupervised for much of the day. His mom couldn't afford to send him to an after-school program, and no one was at home to watch him. This is an example of a disturbing nationwide trend: teens from lower-income families are spending on average over two and a half hours a day *more* on screen time than their wealthier peers. Similarly, teens whose parents have no more than a high school education spend almost two hours more on media than teens of parents with a college education.

That said, it is not simply the *amount* of time on screen that matters. It is also what kids are *doing* on screen. After a ten-year examination of two communities with widely divergent economic resources, researchers Donna Celano and Susan Neuman found (among many other things) that screen-based technologies were *widening* the achievement gap they were supposed to close. Celano and Newman studied the computer use experiences of kids from one of Philadelphia's poorest neighborhoods and one of the richest. Their research showed that children from wealthy families have radically different experiences with technology than their less wealthy peers. The central difference is, perhaps not surprisingly at this point, the role of the parent in the technology usage. That is, young children from wealthy families typically use technology with a parent at hand. In fact, they report that more affluent kids have seventeen times the amount of adult interaction when using technology as do poor kids. This adult attention has everything to do with what a child can get out of the computer or tablet. With parents paying close attention, the wealthy kids in the study spent far less time on games and amusing cat videos than did the less affluent kids. Over time, the benefits of this parental involvement served to widen the gap in student achievement.

The disparity in usage rates can also be seen in studies on race and screen time. African American teens average almost three hours more of media time per day when compared to whites. And while Chet's mother made a commonly used point, that screen time is possibly a better activity than other activities typical of unsupervised teens, it still wasn't helping him achieve academic success.

As the cost of electronics such as cell phones continues to drop, their availability increases. Today, children of lower-income families are just as likely to have cell phones as those from higher-income families. If the time spent on technology created all the learning and benefits the protech supporters claimed (such as the claim that students are using social media and video games to learn), you would expect to see the demographics spending the most time on their phones doing better in school, not worse. But these groups are not doing better. In fact, they are doing worse in every conceivable measure of academic success. An increased availability of technology has not only done nothing to close the achievement gap, it seems to be making it wider, and at an accelerated pace.

Consider this odd trend on the other side of the spectrum. In 2007 Bill Gates admitted to putting strict restrictions on his children's tech use after he noticed his ten-year-old daughter becoming addicted to a video game. When his son complained about the forty-five-minutes-a-day limit, Gates reportedly told him, "When you move away you can set your own screen limits." In 2010, when journalist Nick Bilton asked Steve Jobs how his kids liked the new iPad, Jobs replied, "They haven't used it. We limit how much technology our kids use at home." And the founders of Microsoft and Apple aren't alone. Apparently, many of the tech industries' top executives are putting strict limits on their children's screen time.

What is it these wealthy tech executives know about their own products that their consumers don't? Chris Anderson, former editor of *Wired* magazine and chief executive of a drone tech company, 3D Robotics, described himself and his wife as "fascists" when it came to limits they impose on their children's access to technology, adding, "That's because we have seen the dangers of technology firsthand. I've seen it in myself. I don't want to see that happen to my kids."

This rejection of technology for children has given rise to the popularity of Waldorf schools. These private schools are built on a century-old philosophy that touts the benefits of a hands-on education, by providing in-person social interaction and focusing on creative problem-solving. Consequently, they also reject the use of technology both in and out of school. Some schools go so far as to make the students and their parents sign contracts promising limited exposure to technology even when they're at home. Void of computers, iPads, and other new gadgets, their classrooms are filled with energetic and engaging teachers who design creative, outside-the-box lesson plans. They have chalkboards, No. 2 pencils, and even knitting needles to teach their students. The theory behind the Waldorf philosophy is that teachers should have a great deal of autonomy in how they foster creative and socially competent students. Hands-on teaching encourages students to experience tactile rather than virtual learning. With tuition as high as $24,000 a year or more, however, they aren't necessarily accessible for poor families.

These antitechnology schools have found popularity in an unlikely place: Silicon Valley. Parents who work in high-tech industries find them particularly appealing. Back-to-school nights at Waldorf schools are a who's who of the technology world, with executives from eBay, Google, Yahoo, Apple, and Hewlett-Packard who all choose to send their kids to such schools. Seventy-five percent of Waldorf students in Silicon Valley have ties to the tech industry. Forty of the country's 160 Waldorf schools can be found in California, not including some of the nonaffiliated knockoffs also heavily concentrated in this area.

And with 94 percent of them going on to college, clearly these students are not experiencing any major setbacks resulting from their low-tech education. As Alan Eagle, an executive at Google who is a Waldorf parent and has a computer science degree from Dartmouth, explains, "At Google and all these places, we make technology as brain-dead easy to use as possible. There's no reason why kids can't figure it out when they get older." They take the approach of using early education to cultivate more fundamental skill sets, and only implement technology sparingly.

The fact that a low-tech education can still produce successful students shouldn't be surprising. What is surprising, however, is how the higher-ups in companies that are responsible for putting technology into classrooms are removing their children from the type of educational environment they've worked to create. It's interesting to think that in a modern public school, where kids are being required to use electronic devices like iPads, Steve Jobs's kids would be some of the only kids opted out.

While the kids of many wealthy tech executives are getting a low-tech education, many kids in our lowest socioeconomic levels are getting a high-tech one. At most public schools, the lowest-level ESOL (English for speakers of other languages) students tend to be immigrants from economically underdeveloped nations. Many of these nations don't have the infrastructure for landlines, so cell phones have become a cheap alternative. Because of this, most of these students not only have cell phones of their own, but they are now also provided with their own iPads, tablets, or laptops as part of misguided initiatives to help improve their test scores.

However, these devices have proven to be a bigger distraction than a benefit. Like their wealthier peers, poor students—even those who are illiterate—manage to find unproductive ways to use screen time. One teacher I interviewed says that her ESOL students were only using their school-provided iPads for sending each other pictures of themselves. Once the iPads were abandoned, students went back to using their phones for texting. Fluent in English or not, they quickly learned how to use the voice commands to pull up their favorite movies, television shows, games, and songs.

Not surprisingly, there is a direct correlation between time spent unsupervised and time spent playing on electronics. Students from families in which the parents work multiple jobs spend much of the day unsupervised. This is a larger issue for students from poorer families. The large chunks of unsupervised time lead to students who have grown accustomed to being on a screen for most, if not all, of the day. By the time they get to school, the habit is often so ingrained that it becomes virtually impossible for them to put their phones down.

Now imagine handing out iPads in a classroom riddled with students like Chet, who have these strong cravings owing to excessive nightly gaming. Ed-tech advocates would say that a strong teacher presence in the classroom could prevent misuse of the devices, keeping kids on track and productive. But is this a reasonable expectation for a teacher (or anyone, for that matter)? This would be like handing out full glasses of wine at an AA meeting and asking the leader to make sure the meeting participants were only using them to make music on the rims.

The issues created by screen addiction—video game addiction in particular—have led to the rise of a new achievement gap: the gap between boys and girls. In the last decade, males have seen an increase of high school dropout rates, becoming 30 percent more likely to drop out than their female peers. Their reading scores, college admission rates, and college completion rates continue to decline. Famed psychologist Philip Zimbardo has now dedicated much of his recent work to studying, as he puts it, "the demise of guys." According to Zimbardo, "This demise can be traced to the rise of technology enchantment."

Consider this: psychologists estimate that the average young person will dedicate roughly ten thousand hours to playing video games by the time he or she turns twenty-one. We also know that boys account for the overwhelming majority of video game playing. For context, it takes about 4,800 hours to earn a bachelor's degree. And remember, this is just time spent playing video games, not any of the other digital activities to which a boy may be drawn, such as watching videos of grown men playing video games.

It's the type of digital activities to which boys are drawn that is creating this "demise." In the United States the amount of time spent gaming averages eight times higher for boys than for girls.

Girls tend to spend more time on social media and texting. Along with its more addictive nature, gaming seems to have a more negative cognitive impact than social media. A study published by the American Academy of Pediatrics found that people who spend more time playing video games have more attention problems, and individuals who are more impulsive or have more attention problems subsequently spend more time playing video games. This study also found that video game addicts are twice as likely to have ADD or ADHD. This could help explain the recent explosion of diagnosed attention disorders. A 41 percent increase of ADHD diagnoses in the last decade has disproportionately impacted boys. Now, one out of five boys is diagnosed with ADHD and 85 percent of all attention-related medications are prescribed for boys in the United States. So much for the claims that video games help improve a child's ability to focus.

Besides diminished attention, it seems that video games can have a detrimental impact on memory as well. Dopamine releases, such as the ones intentionally created by video games, trigger the brain to remember the events that created the release in the first place. Memories of these emotion-filled events take priority over the more mundane events that took place earlier in the day. For a child like Chet, who begins his marathon gaming sessions immediately after school, the "more mundane" events being dumped from his memory include everything he's learned in school that day. Instead, he remembers sweet headshots he had in *Call of Duty* or the different types of Pokémon he captured.

The other issue created by extreme gaming is that it prevents needed mental downtime. During periods of lower mental cognition, a person's hippocampus (the portion of the brain critical for memory creation) communicates with the outer cortex (the portion of the brain that stores these memories) to create long-term memories. As neuroscientist Lila Davachi explains, "Your brain is working for you when you're resting, so rest is important for memory and cognitive function." A study supporting this claim showed that students who played video games after a vocabulary lesson remembered far fewer words than those who didn't. In this case, doing nothing after school would be better for Chet than playing video games. When teens neglect rest and sleep by taking their gaming session into the wee hours of the morning, they never give their brains the chance to file away that day's memories. Everything they learned that day in school goes out the window.

According to Zimbardo, video games aren't the only thing wreaking havoc on millennial males. Of course, there is also porn. Although it's difficult to find an accurate estimate of what percentage of the Internet is dedicated to porn, anyone can recognize that a young person's access to pornographic material is greater now than ever before. Today, the average teenage boy watches over seven pornographic video clips a day. Apparently parental blocks and other preventive measures are doing little to stop underage teens from viewing this.

Considering teenage boys' raging levels of hormones coinciding with their unfettered access to mountains of pornographic material provided by their digital devices, it's shocking they even manage to use their devices for sixteen minutes a day for schoolwork. The digital immigrant equivalent of this would be if the teacher stapled a *Playboy* or *Hustler* magazine in the front of every teenage boy's textbook and then their parents sent them off to their room to do their homework on chapter 12. This, however, is precisely what an online textbook or digital assignment is to a digital native boy. In order to even start this assignment, they have to fight off both their drug addiction–like urges to game as well as their most primal hormone-driven sexual urges. Getting to the log-in page of their textbook is a feat in and of itself.

All these factors were stacked against Chet. Unfortunately, this story doesn't have a happy ending. After months of battle between Gabe and Chet on his cell phone use in class, Chet finally lost it. One day, he stood up in the middle of a lesson and started shouting at the student sitting next to him that he was going to "kick his ass" if he didn't give his phone back. Apparently, the other student had taken it as a joke when Chet had his back turned. After Gabe told Chet to go out to the hall so he could cool off, Chet erupted. He charged to the front of the room where Gabe was standing, grabbed a pair of scissors off Gabe's desk, and told him he was going to use them to stab Gabe in the throat.

Needless to say, that was the last day Chet was in class. After a brief suspension he was moved into a smaller class setting in a school reserved for students with emotional disabilities. Several weeks later he was expelled for selling drugs in one of the school's bathrooms. The last Gabe had heard, Chet was in jail. By the time Gabe was telling me this story, he had come to realize that he handled Chet wrong from the very beginning. And although it might not have changed the course of Chet's life, handling his technology use differently certainly could have changed the outcome for his class.

Chet's phone and video games worked to distance him from the adults in his life. To the developing adolescent, a strong role model or mentor is one of the most important things in shaping who they become. Teens from lower-income families already have a higher likelihood of an absentee parent, in particular, a father. This leaves a very important void in their real-world network of relationships.

Technology only works to exacerbate this problem. Underprivileged teens aren't using their devices to forge new relationships with mentors. Consider that the average American teenage boy spends only thirty minutes a week in face-to-face conversation with his father, compared to an average of forty-four hours a week spent on screens. Teenagers are using their devices to replace their real-life parental relationships with video games, movies, porn, and YouTube makeup tutorials. These are what many in this generation are raised on.

This is where, why, and how non-screen-based education is especially important to economically disadvantaged youth. Teachers can become needed mentors for struggling teens. Their real-life relationships provide students with living examples, and they can provide guidance and much-needed support.

Perhaps the saddest thing about all of this is that both sides—those favoring the use of screen-based technology in education and those who oppose it—see the critical need for strong role model relationships in order to help close the achievement gap. It's just that those in favor of screens in school apparently believe that adding screens to the school day is a good way to foster those relationships. It would be wonderful if that were true. It simply is not. Adding screen time to the school day is not what kids need, and those students lagging behind need it less than anyone.

A study done by the Harvard Graduate School of Education illustrates just how influential strong teacher-student relationships can be. Professor Hunter Gehlbach, an educational psychologist, had both students and their teachers fill out a survey about their likes and dislikes on a wide range of topics. Then, for the experimental group, the researchers would sit down with each student and teacher and go over the things they had in common. The idea being that this would strengthen the relationship between the two. After then tracking the students' progress in school, they found that the students with the stronger teacher relationships performed significantly better than the students who hadn't had this simple intervention. The groups with the greatest improvements were African Americans, Latinos, and lower-income students. In fact, it led to an unheard-of 60 percent reduction in the school's achievement gap.

This is just one study of many that support the idea that a stronger relationship between teachers and students improves student performance. This is especially true of students who may lack stronger relationships at home. It illustrates the importance of actual, real-life relationships in human development. Almost anyone who is the product of a traditional education could point to one or more teachers or professors who powerfully influenced them. These aren't the type of relationships that can be replicated by technology. As Paul Thomas, a professor of education at Furman University and expert in public education methodology, says, "Teaching is a human experience."

Even in the hands of a skilled teacher, technology works to separate teachers from students and weaken this relationship. With online classes, students may never even meet their teachers. With flipped classrooms, the teaching is done through the computer as teachers play a secondary role in the classroom. In fact, that's a common mantra of the ed-tech movement. Teachers are relegated to the sideline as students are given more control of their own learning. I've watched many "integrated classrooms" in action, and they almost always involve students silently working on their laptops while the teachers sit in the corner, silently working on theirs. The only direct interaction they have is at the beginning and end of each class when they tell the students to get out and put away their computers.

But this is not what students need—especially students who are struggling. Teachers are more than fact disseminators. They're more than custodians of laptops and iPads. People don't get into the profession of teaching because they're passionate about troubleshooting network connections and uploading YouTube videos to Google Classroom. It's hard to imagine a future where people fondly recall webinars that inspired them, or prerecorded lectures that changed their lives, apps that pushed them to excel when they wanted to give up, and YouTube clips that made them believe in themselves.

This illustrates Gabe's failures with Chet. First and foremost, all young people, especially kids like Chet, need to know they're cared about. They need to know that the adults in their lives have their best interests at heart. By allowing Chet to continue to use his phone in class, Gabe sent him the message that he was uninterested in Chet's success. Gabe wasn't trying to connect with him, or push him to excel. Gabe's actions told Chet that he didn't believe in him and that he wanted Chet to sit there quietly and not disturb any of the other students. What Chet needed wasn't more screen time, he needed a strong mentor who would push him to achieve his full potential. He needed to be inspired. His cell phone only worked to get in the way of that.

This is evident by the success of schools that ban cell phones. The London School of Economics did a comprehensive study of schools that were in the process of adopting a strict no-cell-phone policy. What they found was that banning cell phones led to a 6.4 percent across-the-board increase in test scores. However, the lower-achieving students had the greatest increase, averaging a 14 percent improvement. By removing the screens that were separating the students from their teachers, these schools turned F students into C students.

The gaps in student achievement—rich/poor, male/female, native born/immigrant—are obviously the result of many factors. Technology use in the classroom and at home is clearly one of those. Closing the gaps requires measures by schools, researchers, and policy makers.

We must all become better consumers of information about technology's effects on kids. Perhaps the most frustrating and dangerous part of this issue in general, and the achievement gap in particular, is the way in which otherwise reputable institutions claim that technology is a magic bullet. Stanford University's Graduate School of Education, for example, advertised a report proudly entitled "Technology Can Close Achievement Gaps, Improve Learning." That sounds great. However, the report was actually done by the Alliance for Excellent Education and the Stanford Center for Opportunity Policy in Education. What do these two groups have in common? They are both heavily funded by Bill Gates. Note that this report is not a scientific study, yet it is trumpeted on the website of a graduate school of one of the most prestigious institutions in the world. As with most of the other issues surrounding educational technology, the preponderance of hard science shows troubling effects when it comes to the achievement gap. Reports that show the gap closing are most often funded by those with a financial interest in spreading educational technology.

By contrast, an actual peer-reviewed study by Duke University economist Jacob Vigdor and his team shows that computer technology is associated with "statistically significant and persistently negative impacts on student math and reading test scores." Further, they conclude that "providing universal access to home computers . . . would broaden, rather than narrow, math and reading achievement gaps."

Whenever a study or report shows great things happening with screens in schools, we must look at who bankrolled the study. The bond between school decision makers and the educational-technology industry has become extremely strong. We examine that bond in chapter 9.

**TAKEAWAYS**

- School systems serious about closing the achievement gap should invest more in their teachers and in reducing class sizes and less in technology that only works to stymie human interactions. Doing this will help strengthen student-teacher relationships, something that is proven to increase student performance, especially for the lowest-performing students.
- Supervise your children's technology use. Although parental filters can be used to block inappropriate websites, even the most technologically inept teenagers can become Kevin Mitnick–caliber hackers when something gets between them and their favorite Internet site. Direct parental supervision is invaluable in assuring appropriate and productive use of technology. One parent I interviewed only allows his kids to have flip phones, and he keeps his Wi-Fi router in his briefcase, only plugging it back in when he gets home from work.
- Replace kids' digital time with physical activity. If, as a parent, you can't be home when your child gets out of school, push them to join athletics or other physical extracurricular activities, like marching band. Not only will your children have adult supervision, it will also lessen the time they're on their devices, and the exercise will promote mental growth. Physical activity has benefits across the socioeconomic spectrum.
- Be aware of the signs of technology addiction, which is particularly damaging for kids who are already struggling in school. I've talked with many parents who buy their kids devices like iPads or Kindles with the best

intentions, only to have their kids transform into addicted monsters within a month. Fights over limits become too disruptive, and eventually parents cave in. Remember, almost all the games available today are intentionally designed to be addictive. If by high school, a student has developed interests in healthier pursuits, is active, and doesn't appear to have an addictive personality, limits can be relaxed. However, children younger than high school age really haven't developed strong enough impulse control to fight off the addictive nature of modern technology.

# The Education-
# Industrial Complex

*Screens in Schools Are a $60 Billion Hoax*
—TITLE OF AN AUGUST 2016 *TIME* MAGAZINE ARTICLE
BY DR. NICHOLAS KARDARAS, AUTHOR OF *GLOW KIDS*

IF YOU'RE LIKE MOST AMERICANS, you do not view schools as businesses. Saying "Have a great day at work, honey!" as you watch your nine-year-old walk out the door in the morning is probably not the image you have in your head. Nor do I. Teachers, though, sometimes go a step further. We like to think of schools as almost sacred places—places where students learn the skills and knowledge to help them grow into well-rounded, fulfilled humans who are able to give and get meaning out of life.

While I still *like* to think of them that way, my eyes are now opened. Education is an industry, schools and school systems are customers, and learning—notoriously difficult to quantify—sometimes takes the backseat. Understanding this will help us all as we try to help our kids thrive and successfully navigate the modern school system.

## The Pressure on Decision Makers

As I have said, learning is hard work. Teaching is hard work. Anyone who tells you that they've found a way to make learning and teaching easy is trying to sell you something. In fact, there are many people and companies trying to sell you (and the schools your kids attend) something. So I can be clear at the outset of this discussion, I'm not saying that there is a secret vast conspiracy by industry to make obscene profits by selling families and schools technology that is actually bad for kids. I'm saying that there is a *very open* vast conspiracy by industry to make obscene profits by selling families and schools technology that is actually bad for kids.

On its face, that might sound like an outrageous claim. But think about it. We have already seen many of the ways in which the overuse of educational technology is bad for kids. How could industry profit from that? Put yourself in the shoes of a local education official—a principal, school board member, or superintendent. You likely got into education, like so many of us, to save the world. You learned your content, your kids' needs, and how to navigate your school bureaucracy. You learned about your school community and how to articulate a vision for the students you wanted to lead. Your vision likely (I hope) included something about preparing kids for the future. A half-century ago, if you worked hard and were competent, you were probably able to rise through the ranks, lead your school or system for a decade or so, work on your vision, retire, get your gold watch, and enjoy a nice long cruise into the sunset thanks to your pension. The school or system next to yours likely had someone or some folks at the helm with similar stories, similar goals, and similar gold watches. The nation was filled with these stories, a patchwork quilt of world savers with visions and gold watches. Some succeeded, others failed, and the system chugged on. I'm not saying that was right or wrong. It just was.

However, in the last few decades there has been a shift. Schools and school systems are far less able to operate in isolation. Like anything else, this has costs and benefits. Schools and those who lead and teach in them are now more accountable. But accountable to whom? In the story above, the benevolent retiree might have been accountable only to him or herself, or to the school board who hired him or her. Now, however, leaders of schools are accountable not only to themselves and their bosses but also to parents, unions, and decision makers at the local, state, and federal levels. For most of our history, education was only a local issue. That has been turned on its head, and education is now a staple of presidential campaigns and actions. In 1994, for example, President Clinton signed the Goals 2000: Educate America Act with the help of a Democratic Congress. At that time, and until the year 2000, the Republican Party platform called for the abolition of the US Department of Education. The parties stood in stark contrast on this issue. But in 2002 Republican president George W. Bush signed into law the No Child Left Behind Act. This left the United States with two major parties both believing that education was an issue for the federal government. Pity our poor local school official.

I take no position on either of these acts, or even on the role of the federal government in education. However, what is clear is that these steps marked an enormous increase in pressure on schools. Failure was not going to be tolerated (which, of course, it shouldn't be). Failing schools were brought to the consciousness of a nation anxious about how its children were going to compete with students from around the world—children who seemed to be getting smarter faster than ours. People have always been concerned about schools, and likely always will be. In the past, though, concern might have meant a PTA meeting in a school gymnasium at which the superintendent was asked tough questions by worried parents. Now, politicians on both sides of the aisle in local, state, and federal offices scream for change and call for the heads of those in charge.

So if you're in charge and the nation is calling for your ouster, you better articulate a vision for the future. You better have a plan to bring our children the skills they need to compete. You better understand that things are different now, the old system is broken, and you must make changes. Whether or not you believe the old system is broken or that change is necessary, you likely won't keep your job if you don't *do something*. Enter industry.

## The Open Conspiracy

In 1994, the same year President Clinton signed into law the Goals 2000 Act, Dr. Anthony Picciano warned of what he called "the education-industrial complex," clearly echoing Dwight Eisenhower's warning about a military-industrial complex. He has since coauthored an excellent book on the subject that explores this topic in much greater detail than I will here. Dr. Picciano was forward thinking enough to see what was going to happen: federal executive branch involvement in education would lead to legislators demanding better schools, school officials being hungry to satisfy the public's demand for those better schools, and industry being hungry to profit off these demands. More prescient he could not have been.

To wit, a few years ago Rupert Murdoch of News Corp. said that educational technology is a "$500 billion market that's largely been untapped." This makes the $360 million his company spent purchasing Wireless Generation (an educational-technology firm) seem like a teacher's salary. Steve Jobs believed that online

textbooks were "an $8 billion market ripe for 'digital destruc-
tion.'" *Fortune* magazine, in a 2015 article questioning the wisdom
of educational technology in the classroom, cites a forecast that
spending on digital devices for classrooms will near $20 billion by
2019. In 2013, US schools spent $4 billion just on mobile devices.

THE *Journal* has been covering educational technology since
1972. *THE* used to stand for *Technological Horizons in Education*. They dropped that several years ago, though it is still an
educational-technology journal. *THE Journal* reports that nearly
90 percent of US school districts plan on acquiring at least some
tablets for students during the 2016–2017 school year. The percentage of school districts planning these purchases is highest
for the youngest students. *THE Journal* got this information after
reviewing EdNET Insight's report "State of the K–12 Market 2015."
If you're interested, you can buy the report for $7,500. I'm serious.
At any rate, *THE Journal* includes articles with titles like "The Art
of Procurement: Balancing Price and Performance" and "Learning
Management System Market Expected to Grow $10.5 Billion in
Next 5 Years." The focus is directed entirely at growing corporate
profits by selling educational technology to schools. There is little
discussion about what is best for kids. This is part of what I mean
by the conspiracy being very open.

Each fall, in fact, there is an EdNET conference somewhere
in the United States. This conference is described as "a must-
attend event for education industry stakeholders." Education is an
industry, like chemicals, airplane engines, or, um, computers. The
purposes of the conference are the following: (1) make immediate
connections; (2) forge new business connections through productive networking; (3) gain essential market insight; (4) build your
list of receptive market contacts and potential partners for growth;
and (5) determine the right solutions for your business based on
current market trends. You'll likely notice that there is not one

word about teaching kids or making schools better. The purpose is to figure out how to make as much money as possible on the sale of educational technology—whether that technology is needed or not. As of the publication of this book, over 150 companies were signed up to attend the 2016 conference. The superintendents of a few school systems were also there as keynote speakers. However, since the 2016 registration fee ranged from $1,395 to $1,695 for the three-day conference (not including travel or hotel), and since it is during the week in the fall, the target audience for this conference is clearly not the people who will be trying to use all this technology to teach your kids how to read and write.

Of course, there is nothing wrong with companies working hard to make profits. One of the subjects I teach is economics. Markets are beautiful. I love capitalism. However, we need to think hard about profits earned by selling schools products that make it harder for kids to learn. When private companies are preying on naive and uneducated consumers, someone needs to step in. We know ed-tech firms are not going to stop selling—nor should it be their responsibility to give up their bread and butter. We know the local school officials, whose areas of expertise typically do not include marketing, business, and technology, are not in a position to say no. That leaves government to intervene.

With all of the recent federal involvement in education it would be natural to expect the US Department of Education to have a say in this. They sure do. According to their National Education Technology Plan, *Transforming American Education: Learning Powered by Technology*, the department calls for "applying the advanced technologies used in our daily personal and professional lives to our entire education system to improve student learning, accelerate and scale up the adoption of effective practices, and use data and information for continuous improvement." Think about that. There are over fifty million school-aged children in the

United States. All of them are potential consumers of educational technology, and the US Department of Education wants schools to "scale up the adoption" of these technologies.

So there it is. Schools are under increasing scrutiny, the US Department of Education is pushing for more technology in the classroom, and industry is brimming with companies eager to provide it. One of the most troubling aspects of this is highlighted in that statement from the National Education Technology Plan (the architect of which is a regular speaker at the EdNET conference). The statement says that these advanced technologies will "improve student learning." But we already know that's not the case. In fact, a recent study, "How We Learn" by researchers for *Scientific American Mind*, "reviewed more than 700 scientific articles on ten common learning techniques to identify the most advantageous ways to study." Of those deemed most advantageous for learning, exactly zero used any sort of advanced digital technology. Worse than that, and as I have taken great pains to describe, the overuse of these technologies actually harms kids in many ways. Since no reasonable person seeks to harm children, something else must be going on.

The answer, at least in part, is in the widely held perception that the way in which students learn is changing. We know that's not true, so why would anyone tell us otherwise? If you have not seen the musical *The Music Man* you should, because it's great. It also provides us with an excellent parallel to what is happening in education right now. In the musical, Harold Hill is a charlatan who makes his living selling band instruments and uniforms to folks with the promise that he will teach them how to play, even though he is neither a musician nor a music teacher. He simply sells the goods, earns the commission, and then skips town without teaching anyone anything. In this story, Mr. Hill ends up in River City, Iowa, a town that is doing just fine, and they have no interest in his band ideas. In order to make money, he has to convince people their town is in desperate need of a band to "keep the young ones moral after school." The threat he points to is the new pool hall in town. Once he convinces the townsfolk that pool is evil, they come clamoring to him for all his band supplies because they've "got trouble with a capital T and that rhymes with P and that stands for pool!" Cha-ching.

Playing the part of Harold Hill in our real-life production is the ed-tech industry, educational technology is in the place of the band instruments, the schools are the townsfolk, and the fact that students now learn differently is the pool hall. That is, ed-tech firms created this need for educational technology by selling us on the idea that kids learn differently now. Dell's education website, for instance, says, "The way students learn is fundamentally changing. . . . Dell provides educational technology solutions, powered by Intel, that support this next-generation learning environment and drive successful student outcomes." Apple's education website proclaims, "We've been proud to work alongside educators and students to reinvent what it means to teach and learn." Blackboard has a two-minute video entitled "The Voice of the Active Learner" in which today's student, who is a digital native, explains that schools have to keep up with her. To learn, this student looks "online, because the classroom isn't enough." Every teacher I know has seen this video in staff development training. The Pearson Education website says, "The shift to digital education is well underway and it's changing the way students learn." Companies such as Blackboard, Apple, Dell, Pearson, and many others stand to profit a tremendous amount by convincing the public and school systems that students today cannot learn without advanced, digital technologies. Since we know that's not the case, we've got trouble with a capital T.

I attended a workshop recently at a school that had just adopted one-to-one education (where every student gets a computer). The meeting was filled with the protechnology claims mentioned above. The first speaker opened with, "Technology today has revolutionized the world. Now, my phone can give me directions to virtually anywhere. My car communicates with tollbooths, paying my tolls without me having to stop my car. We are living in a very different world where today's students learn differently." The problems with this logic are many. The claim that several minor increases of convenience have "revolutionized" the world is a bit of an exaggeration. Discovering agriculture, which allowed people to move from hunter-gatherer clans to full-blown civilizations, is worthy of the title "revolutionary." Making it so people don't have to throw quarters in tollbooths is not. More important, how do any of the changes mentioned affect how the brain processes and stores information? Living in a world where cell phones make it easier to get around is not necessarily linked to the idea that children learn differently. What parent has ever driven through a tollbooth and then thought to herself, *That just revolutionized the way my kid's brain works*? Unfortunately, tech advocates like this speaker often simply repeat their illogical conclusions until everyone accepts them as truth.

This revelation about students now learning differently would be fascinating and remarkable if it were true. As we have seen, it is not. To be certain, the way they use and interact with technology is changing. As much as these firms would like us all to believe that gadgets are the long-sought quick fix for education, and as nice as it would be if it were true, that is simply fantasy. Presented with a threat to their children, though, people understandably want a solution, and they want it now. It was true in River City, and it's true in our world, from Bedford, Virginia, to Helena, Montana, from New York to Los Angeles.

Speaking of Los Angeles, the unified school district there in 2013 awarded a massive contract to Pearson and Apple. The deal was intended to provide every one of the 640,000 students in the nation's second-largest school system with an iPad loaded with Pearson software. It would have cost hundreds of millions of dollars—just for the iPads and software. However, the school system's infrastructure was in no way prepared to have all those students logged in to their network at the same time. This required an enormous overhaul of wiring in the schools. The rewiring effort required an additional $800 million. The total price tag was $1.3 billion. Amid allegations of deal fixing and a noncompetitive bidding process, the contract was scrapped, and the bidding reopened. But consider that again—by making *one* sale, Apple and Pearson were set to earn over $1 billion. The device purchases will now be made from Apple as well as Google, Dell, Lenovo, and Microsoft. It is no wonder that corporations are working hard to convince schools that student learning today depends on screen time.

Perhaps the most troubling part of the entire L.A. schools' debacle was a remark from then-superintendent John Deasy when he was explaining his view of the need for iPads in the classroom. He said, "I'm not going to be interested in looking at third-graders and saying, 'Sorry, this is the year you don't learn to read.'" Who would ever want to say that to any child? More important, though, since when is an iPad necessary for a child to learn how to read? Mr. Deasy had taken the reins of a school system in bad shape. His bosses and the public wanted results and they wanted them *now*. The pressure to *do something* must have been overwhelming. It is in conditions like these that we tend to make our worst decisions.

That situation, while expensive and troubling, pales in comparison with the way the Common Core State Standards came into being. Calling them state standards (as opposed to national standards) would be hilarious were it not so misleading. In the military-industrial complex I referenced earlier, decision making is centralized. If you're a defense contractor and you want to build a new jet, for instance, you need to convince a few people in the Pentagon and you're good to go. Education isn't like that. There are over thirteen thousand school systems in the United States, all of which have different needs and priorities. As Dr. Picciano explained to us, that's an awful lot of sales pitches to have to make. What would simplify that? National standards. Having those would mean that all school systems throughout the country would need to teach the same things. If you had products that prepackaged these new standards and content, they would sell themselves. Schools would *come to you* to buy these products. Cha-ching again!

In 2008 Gene Wilhoit and David Coleman approached Bill Gates and pitched the idea for national K–12 math and reading standards. Wilhoit was the director of a national group of state superintendents of education. Coleman was the cofounder and director of a nonprofit group called Student Achievement Partners. He had previously founded the Grow Network, which analyzed test scores for states and large school districts. The Grow Network was purchased by publishing giant McGraw-Hill for an undisclosed amount. Coleman moved on to become president of the College Board, best known for designing and administering the SAT and AP exams.

Why would they seek out Bill Gates? Mr. Gates is brilliant, but he also happens to be one of the world's richest people, and his charitable foundation gives hundreds of millions of dollars to educational causes. Much of the money in this case went to convince politicians and state boards of education that the Common Core standards were an improvement over existing state standards, and the new model was needed to even the playing field nationally. For example, over $15 million went to lobbying groups in Kentucky to get that state to sign on. A single North Carolina group, the Hunt Institute, received $5 million. This group created "sample letters to the editor, op-ed pieces that could be tailored to individuals depending on whether they were teachers, parents, business executives or civil rights leaders." All told, the Gates Foundation spent over $170 million nationwide to get states to sign on to Common Core. Former US secretary of education Arne Duncan, who had been head of Chicago Public Schools, worked to push the standards through. While in Chicago, Duncan received $20 million from Gates to reform high schools there. Ironically, while serving in DC, Duncan's children attended public school in Virginia, one of four states to not adopt Common Core. Forty-six states signed on. Eight have since abolished the program. Currently, the future of Common Core is murky. The effects of the program, though, are long lasting.

Standards, of course, are vital to any process. Common Core might be great. It might be awful. I happen to teach in a state that never adopted the standards, so I cannot speak with any authority on that. However, $170 million is a lot of money. I started to wonder why it was worth so much to Gates and others who lined up behind the standards. It might be that they all really want K–12 students in the United States to be learning the same things. For my purposes, I wanted to know if this had anything specifically to do with educational technology. In a word, yes. Common Core standards come preloaded on the Microsoft tablets marketed to school systems. Publishing giants such as Pearson, McGraw-Hill/StudySync, and Houghton Mifflin Harcourt are busily partnering with firms to create apps that prepackage state standards on their own devices. The online testing and analysis designed to assess how well students have mastered Common Core standards provides many firms, Pearson especially, with a near-endless revenue stream, as does the maintenance, repair, and replacement of devices, the wiring and rewiring of schools, and the teacher training. If anything, Mr. Murdoch may have lowballed how big the industry could be.

But, ultimately, so what? Companies make money off needs. This is not only OK; it is what we want. Incentive generates innovation, and innovation increases our standard of living. I am fine with the fact that firms make money off schools. That has been going on since there have been firms and schools. Someone has to sell schools chalk, markers, and paper, and those people have to be paid. The difference with this, though, is that this particular group of products and services is bad for kids. According to a University of Nebraska study, classrooms in the United States have more technology available than any other nation on Earth—more than Japan, and far more than China. Yet, when compared internationally, what nations do we trail? These and many others—all of which have less available educational technology.

## Other Issues (or More Trouble with a Capital *T*)

In addition to the problems we have already outlined, the relationship between schools and educational technology firms creates a whole host of other issues. One is privacy concerns. For example, as of 2014, thirty million US students, faculty, and staff were using Google Apps for Education. Google is particularly appealing for many cash-strapped school systems because many of their education services are free. Many systems can't afford to *not* use their products. Students are given access to free e-mail, free storage space for their files, free means of collaboration, free GPS maps, and free software. This is why they've gone from a mere 1 percent of the education market in the United States in 2012 to 51 percent in 2015.

But altruism alone doesn't make you $16.8 billion a year and the eighth most profitable Fortune 500 company. Even with giving away almost all their products, Google still has plenty of opportunity for profit. Their education suite offers millions of opportunities for what is known as "data mining." They track and compile information from what students are doing while online, what they're searching in their search engine, what they're writing about in their e-mails, and what they're putting into their documents saved on Google's database. Scarier still, the GPS data allows Google to physically track their every movement. With all this information, Google knows more about their users than their users' own mothers. This information is compiled into a very accurate and up-to-date dossier for each of the millions of children using their products, and then sold to marketing firms. Despite several lawsuits from student groups arguing this violates their privacy, Google still insists they've broken no laws. Most current privacy laws were written in the 1970s, before the widespread sharing of electronically gathered information was possible, and there is typically no requirement for schools to get parental consent to share student information with contractors. In many schools, it's impossible for students to opt out of the Google system. As your student is doing homework, more and more of which requires screen use, a data profile is being constantly created and updated for Google and other companies. What they do with that data is out of your hands.

As of the publication of this book, one of the latest and greatest educational software tools is called Summit Basecamp. At this point, it has a relatively small market share but it is growing rapidly. It was built with help from Facebook. While the software is tracking student progress, it is also collecting data on each student and sharing it with firms, principally Facebook and Google. Student data shared includes names and e-mail addresses but also more personal information such as grades and what sites students visit on the Internet. While this requires parental consent, it is an all-or-nothing program. That is, if your child's school uses the software you'll typically get to decide whether or not to consent to your child using it. However, you cannot just opt out of the data collection. Your child either uses the software (and surrenders his or her data) or not. If you don't consent, though, your child won't be using the same software as every other student in school. The pressure to conform is enormous. As a result, the mined data will continue to grow.

Another issue is that many of the gadgets and apps schools use are either filled with glitches or fail to offer the functionality that supposedly made them superior to analog materials in the first place. Inadequate network infrastructure and device glitches were issues before the impropriety surfaced in the L.A. schools iPad fiasco. In any school, getting students logged in to the network and on to the correct application obviously takes significantly more valuable class time than opening a book or taking out some paper. In terms of functionality issues, here is an example from a nearby school district: It is 2017. George W. Bush is the president of the United States. William Rehnquist is chief justice of the Supreme Court. Colin Powell is the secretary of state. These and many other incorrect tidbits of information can be found in the online US government textbook students in this district currently use. When this online textbook was adopted in 2010, all teachers in that district were sent to a daylong training on how to use it. As teachers browsed the pages, some raised the point that the above officials were already several years out of date. In fact, this "new and improved" digital textbook was more out of date than the older paper version the schools had been using. They were assured that they were looking at a beta version and the beauty of the online textbook was that it could be easily updated, saving the school system millions of dollars when compared with buying new editions of hard copy versions. Seven years later, they were still waiting for the first update, and Colin Powell was still the secretary of state. Online textbook publishers often sell book licenses to schools with the promise that the textbooks will be regularly updated. Based on our survey, this promise is rarely kept. In and of itself, outdated information in an online text is no more of a problem than it is with a hard copy book. These digital books often have none of the other functionality promised, such as interactive lessons, embedded videos, and highlighting capabilities. The product often pushed out is just a PDF copy of the book that is housed on a server that regularly crashes and that is not available for any type of e-reader. The only advantage it offers over traditional books is that it can read passages aloud in a robot voice. Every year since its adoption in the district, teacher e-mail inboxes have been flooded with e-mails from students and their parents complaining about the online textbook. Some parents have dished out hundreds of dollars to buy their own paper versions of the book so their kids don't have to try and read from a PDF all night.

With the online book universally despised by students, teachers, and parents alike, it's a wonder why schools would have adopted it. When teachers in this district asked about the switch, the decision makers gave the typical protech response, "It's the wave of the future and how digital natives learn." But that wasn't true. Teachers had to listen to digital natives complain about the digital technology forced on them, and beg for paper versions.

The reality is digital textbooks aren't better for students; they're cheaper. That's why publishers are pushing them, and that's why school systems are buying them. Textbook publishers no longer have to actually print books. Think of all the overhead they save by no longer having to make the thing they sell. All they need now is an intern armed with a scanner and they can produce one file that can be sold hundreds of thousands of times. Schools can rationalize this cheaper alternative by saying things like, "This is the wave of the future," and, "This is how kids learn today." They do this because it sounds better than the truth, which is, "We're going to give your kids books they're going to struggle to read because it'll save us a lot of money."

But rarely do these cheaper alternatives actually save money. Rather than admit they wasted millions of tax payers' dollars, schools doubled down, and invested more in technology. Complaints to the textbook publisher in the previous example about the lacking functionality were met with an offer to buy an upgraded version that offered some of the things promised. When policy makers realized that teachers didn't have access to digital textbooks in their classrooms, schools had to buy tablets and laptops for teachers to share. When there weren't enough, schools had to buy more. When there still weren't enough, schools bought every student a laptop.

Further, giving children expensive, delicate devices seems like an odd choice. I know as a curious boy I would certainly have wondered how far an iPad could fly if thrown like a Frisbee. When was the last time you expected your eight-year-old to take care of a $500 piece of equipment every day? The child may well be overwhelmed with the stress of not breaking it, or they may get a tad careless (as children are wont to do), or both. In any event, devices get broken, and schools must pay to replace them. Hoboken, New Jersey, for example scrapped their laptop-for-every-student program when the repairs and maintenance became unsustainable. "Screens cracked. Batteries died. Keys popped off. Viruses attacked," wrote *Hechinger Report*'s Jill Barshay when reporting on the situation. This was all happening to laptops with reinforced hard-shell cases, since the school system knew that kids would be tough on computers. In 2013 Quaker Valley High School purchased 1,173 laptops for its students. After one year, 490 of them had to be replaced. There are dozens of other examples like this. When a student loses a textbook, it is typically under a hundred dollars to replace, and that money typically comes from the student. Most school systems in most circumstances, though, will not charge students to replace lost or damaged laptops, as they are too expensive. That cost falls on the school and, ultimately, the taxpayer. Where does that money go? Ed-tech firms.

So, what can parents and communities possibly do about this? Quite a lot, it turns out. However, the battle will not be easy. Billions of dollars are at stake. By the time your child gets to school, he or she will have been the target of an aggressive educational-technology marketing campaign for years. CTA Digital, for instance, produces something horrendous called the iPotty. It is a potty-training toilet that has an arm that will hold an iPad or other tablets, so the young child does not even have to be disconnected from the screen when using the toilet. It is selling nicely. It gets worse. You know Fisher-Price, right? That's the company that produces all sorts of wonderful things for little kids. They're owned by Mattel. Mattel is the company that produces many toy classics, such as Matchbox cars, Barbies, and Rock 'Em Sock 'Em Robots. They also produce something almost unfathomable called the "Newborn-to-Toddler Apptivity Seat." This is an infant seat that keeps the iPad squarely in the baby's field of vision. Think about that. Before the infant can control her head, she can sit in a seat that will keep a screen front and center. This targeting of children is perhaps best exemplified by ABCmouse.com. This, and other sites like it, tell parents that their children will be behind if they are not plugged in and online from a very early age. By the time they get to school, then, children are haggard veterans of the battle with the education-industrial complex.

The tide can be turned, but it will take a while. There is only one way the codependent relationship between schools and ed-tech companies is going to end: these companies have to stop making money off schools. That is not going to happen by itself. It will only end if schools have a change of heart, look at actual science, and stop selling our children's brains to the ed-tech industry. It is therefore imperative that all of us who understand what is at stake confront schools and school decision makers with what we know.

## TAKEAWAYS

- Opt children out of screen time whenever possible. Schools will not make this easy. The one-to-one program sweeping the nation, wherein each child gets his or her own device to have at school and at home, makes this a tough course of action. But it can be done. Some school systems are more willing to listen to parents' concerns than others. But all should have an opt-out option for programs like these, or any others in which students are required to be on screens. All public schools certainly have the option for students to *use* a computer in class to take notes (even though research shows such note taking to be less effective than handwriting notes). They should, therefore, have the option for a student to *not use* a computer. In some counties, students may opt out of one-to-one and not have a device assigned. However, the child is typically not given the option of opting out of the computer-based instruction. Therefore, it is incumbent upon the child, each day, to check out a device for use during the school day and then return it after school. Depending on the size of the school and the number of students opting out, this can be a painful, time-consuming option for kids. However, if enough parents start opting kids out, schools will have to address the issue—or have hundreds of kids late for the start of the school day as they wait in line to check out machines. To wit, Dr. Victoria Dunckley, in her outstanding book *Reset Your Child's Brain*, recommends a "screen fast" for kids in order to get their brains back on track. She says that one of the thorniest obstacles to overcome should one choose this option is how to handle schools. Initial resistance from schools, however, should not stop parents from advocating for what is best for their kids.

- Confront schools and school decision makers with what you now know. If you simply say you want to opt your child out of one-to-one, you could be branded as "that parent" who is a pain in the behind. That can come back to haunt your kid. However, if you can give schools a reasoned, dispassionate, cogent explanation as to why you are opting your son or daughter out, you will be received much more fairly—and you might just change a few minds. Remember that you're on the right side of this issue. You have actual science to back you up. Schools do not. In fact, when I interviewed pro-ed-tech decision makers and explained what I have found, I asked to see the science they were using. It took a while. After a week of waiting, I was finally referred to Project RED. *RED* stands for "revolutionizing education." Their website describes this as "ongoing ed-tech studies." This is intended to show the research behind the one-to-one computing movement. A few more clicks on their website leads you to the sponsors of their "research." Guess who: Intel, Dell, Pearson, Hewlett-Packard, and SMART Technologies. Not surprisingly, the research conducted by these technology companies found that more computer time in school is great for kids. Specifically, technology in school is most effective when "technology is integrated into every intervention class period," and "students use technology daily for online collaboration (games/simulations and social media)." Wolf, meet henhouse. "Studies" by tobacco companies in the 1950s and '60s showed that cigarette smoking was good for you, or at least not as bad as medical science was showing. Actual impartial science was pretty much unanimous: smoking is bad for you. We are in a similar position today. Unfortunately, educational policy makers are taking their bait hook, line, and sinker. We cannot forget this when advocating for a return to sanity in the schools.

- Work with your school's parent-teacher organization to raise the issue and at least begin this discussion. Be aware, though, that plenty of other parents have bought in to the claims of ed-tech firms. You may find many different parent groups unwilling to even have a discussion focusing on the negatives of technology. In a conversation I had with a head of one such parent organization, I was told, "We have no way of monitoring the use of technology amongst the students once they are outside of school. . . . Our children are the ones teaching us, the adults, how to stay up to date with all these new advances." Notice that first statement—parents now have no control over their children's use of technology outside school. Essentially, this person said, ed-tech is here to stay so we better get on board. But the last part is puzzling: if students are the ones "teaching us the adults" about "these new advances," then why should we attempt to teach these new advances in school in the first place? This type of confusion needs to be sorted out. The more we speak up at PTA meetings and other public forums, the more school communities will at least engage in discussion about the issue.

- Share what you know with the schools. There has always been some tension between schools and parents, and likely always will be. That's to be expected, and it can lead to productive outcomes for both sides. However, the tension typically exists about a particular parent's particular child. Conflict is typically more immediate and targeted. "My son deserves a better grade on this project." "My daughter should have the chance to retake that test." "I know my son didn't plagiarize that paper because I wrote it." (True story.) Parents don't often complain to a school about the school's mission statement or general direction. What we don't often hear in schools is "The set of tools and techniques you are using as a school is bad for my child." That's because parents have their own jobs and duties and preoccupations. Most parents assume that the school is doing the best it can with the resources it has. Sure, Mrs. Schlonagel is too inflexible in English class and Mr. Garcia's math class is extremely boring. But overall the school is doing its best. We all want that to be true. However, parents need to have the facts, and the central fact is that schools are pushing more screen time on kids, and that's bad. There is no harm in sharing information about this. Don't be shy.
- Take it up a notch. Once you have a critical mass in your PTA at least demanding more study on the issue, take it to the school board. Every school board I researched has some way for the public to comment on policy. If you speak alone and ask that schools stop purchases of screens for kids, you may well be seen as the one wing-nut Luddite who wants to stop kids from learning.

However, if your PTA shows up and speaks with one voice, a board will have to take notice. Better yet, if you connect PTAs from different schools and show up together, or speak one right after another, the conversation will have begun and you will then have the reins.

- Learn from the success of other issue-advocacy groups in schools. A major issue facing education is sleep time for teenagers. In districts across the country, small groups of concerned parents have advocated for later daily start times for high school students. These groups of a very few, very loud, committed parents have demanded that school boards look into the issue. And school boards are listening. Many high schools across the nation are opening later in the day. Parents, teachers, and students are still split on this issue. However, it is getting done because small groups are getting organized and demanding research into the issue.

- Take a page from the playbook of the other side. Remember the Hunt Institute from earlier in this chapter? They created letters to the editor, op-ed pieces, and talking points for people to advocate and articulate their position. People are busy. Asking them to hear your ideas is one thing. Asking them to research and write letters is quite another. However, if you have prewritten letters for folks to send to their local papers, school boards, and schools, they will be much more likely to participate. They can simply sign the letters and send them as they are, or you can send electronic versions they can tailor to their specific situations.

Again, none of these ideas is a quick fix. Change in schools requires concerted, determined efforts by a large number of people, starting with you. The thing to keep in mind, though, is that your first concern—I assume—is your own kids. You don't have to change US education policy. Just work on the schools your children attend. Bit by bit we can change the momentum. Again, companies will only stop selling to schools when schools stop wanting more screens. The more we can work together to help schools see what is happening, the more we can create the schools that will best serve our children. My vision for that is in the final chapter.

# 10

# Ideal Education
# in a Modern World

*Teaching simpler and learning simpler.*
—A NEW MANTRA FOR EDUCATION OF THE FUTURE

HERE IS A PEEK behind the curtain: teachers love to be loved. We love it when students are enraptured with our lessons. We love the feeling of students being engaged with the material we present, and the hosannas rain down. In this way, we are egomaniacs. As such, we all have that one lesson we cannot wait to teach every year. It is the lesson that has the students so engaged that they are shocked and sad when the bell rings. You actually hear "Awwwwww . . ." when the period ends. Picture teaching that lesson. If you are not an educator, think about your own time in school. Picture a lesson you found inspiring, informative, important, and engaging. Think about a lesson that left you sad when it was over. Then, educator or not, answer this question: How much advanced technology was involved in that lesson? If you are like the overwhelming majority of people I surveyed, the answer is little to none.

Now answer a second question: What made (or makes) that lesson so great? Again, if you are like the overwhelming majority, you will answer something about how important the teacher was to that lesson. The experience of creating a rich learning experience for students has always been dependent upon a well-trained, well-prepared educator. These teachers know the best lessons draw on the natural needs and skills of themselves and their students. Simply giving teachers iPads and other mental pacifiers for use in the classroom does not make for better teachers. A bad teacher with a classroom set of Kindles is still a bad teacher. Similarly, removing technology from a classroom does not make for better teachers. A bad teacher who gets rid of all the classroom laptops is still a bad teacher. I do not necessarily advocate technology-free education. Further, I do not simply want to return to "the good ol' days" when teachers only lectured and could whack students with rulers. Salty, curmudgeonly teachers who are looking for support for the "sage on the stage" model of teaching, which entails lecturing from bell to bell every class, are going to have to look elsewhere.

## Three Guiding Principles

Making schools the best they can be requires more than removing technology that proves to be harmful to educational goals. We should want much more than that. I advocate developing a new way of thinking about education that requires using and reorganizing "old-fashioned" ideas about learning, critical thinking, and social interaction. We have experienced vast improvements in technology in a relatively miniscule amount of time in the human experience. This is undeniable. What has remained almost completely unchanged since Paleolithic times is our basic anatomy. How our brains function, how we develop thinking skills, how we learn, how we interact with one another has been ingrained in all of us. What I advocate, and know to be successful, is a recapturing of our natural abilities: teaching simpler and learning simpler.

This model is based on three core principles: (1) deliver instruction in the simplest possible manner; (2) focus instruction on what students are able to *do*; and (3) foster face-to-face human interaction and opportunities for community building. I call these the three S's of education: Simple, Skills, Social. These principles should guide us today because they use the innate learning abilities and tendencies that all students have. An ideal lesson can be done in any classroom, from the wealthiest school district to the poorest, from preschool on up. It could be done in a classroom, but also on a bus during a field trip, or on the side of the road if the bus breaks down, far from electrical outlets or the nearest Wi-Fi hotspot. Obviously, not all lessons are going to be "ideal." However, these principles should guide schools' planning and instruction. Let's take a look at each of these principles in greater detail.

## Principle 1: Keep It Simple

Do you know what does not have to boot up or find a network connection? Chalk. Do you know what never needs to be recalibrated? A blackboard. Do you know what kind of textbook never has issues loading? A paper one. Do you know what technologies virtually all students can use without explanation? Paper, pens, scissors, and markers. Today's classroom is becoming too complicated. Technology provides too many obstacles to overcome in order to provide instruction. There are too many distractions vying for the limited focus of today's digiLearners. Lessons involve too many variables. All of these complicate the teacher's primary mission, to educate.

Ideal education is about removing the obstacles placed in between the teacher and the student. Instead of the teacher putting a laptop or iPad in front of students in the hopes that the hyperlinks or apps they have provided will do the instructing, I advocate teachers delivering instruction themselves. Teachers can do many things educational-app creators can only dream of doing, and those things are central to education.

For instance, teachers can stop instruction to answer questions. Teachers can gauge by students' body language whether or not they comprehend the material. If students are struggling to understand, teachers can explain the concept differently. Teachers can immediately adapt and differentiate their lessons from class period to class period. Most important, teachers can connect with students on a personal level and develop important community-strengthening bonds with them.

Again, the point is not to remove all technology from the classroom. It is to *simplify instruction.* No reasonable teacher would ask a group of kindergarten students to "extract from thine escritoire a single graphite-based chirography implement" if she meant "get a pencil from your desk." However, we are doing this very thing to our students and schools. An excellent teacher I interviewed, call him Peter, related a story in which he overcomplicated a lesson and then improved it by simplifying it. He decided he was going to use his school's educational-technology tools to jazz up a lesson both he and his students had always found tedious and boring. He spent six hours creating a very good thirty-minute lesson that involved a PowerPoint presentation of period political cartoons, and he reserved a mobile laptop cart for students to use. During the presentation of the lesson he would project the images and give students some background on each. He had also woven in a Google Doc and would ask students to comment on the cartoons.

Their comments and names would appear on the screen, and then they could discuss each cartoon as a class.

The first period he did the lesson he was excited at how well everything went. He realized that because of the Google Doc component, he was getting comments from students who ordinarily do not participate in class discussions. The second class also had a good experience with the lesson. At the end of the day, however, he spoke with another teacher of the same subject (I'll call her Molly). She was doing a similar lesson, using cartoons from the same time period. She had been doing the lesson for years and said it was one of the best lessons she did all year. However, her version was an "old-school" gallery-walk lesson. In short, she had printed copies of the cartoons and affixed them to large sheets of newsprint she had posted around the room. At her prompt, students got up and walked around the room, reading the cartoons and making curriculum-related comments on the newsprint. This lesson took her about fifteen minutes to prepare. Peter decided to alter his lesson to imitate Molly's and see if there was a difference.

The difference was that he took a good lesson and made it great. The simple act of having students get out of their chairs served to energize the classroom and the discussion. Having students walk from cartoon to cartoon, as opposed to having the cartoons switched for them on a screen, meant they had to interact with one another. As he walked around the room, Peter noticed there was content-related discussion happening as the students were writing. When he had students commenting on Google Docs, the room was silent and students were intently staring at their screens.

By simplifying this lesson, though, he had added a dynamic component that was missing. This lesson was now hitting the kinesthetic learners. It was also still drawing commentary from students who do not typically comment in class discussions. It enabled Peter to stand near a cartoon and interact with a small group of students who were working on it. By simplifying the lesson, Peter had increased its value and also made his own life easier. More students were engaged, and they were more engaged than before. Informal assessment was easier. In short, this lesson was superior in nearly every way compared to the more complex technology-enhanced lesson.

The advantage of a simplified lesson is its flexibility. Any experienced teacher knows that any number of unforeseen circumstances can force a teacher to change plans. Once a lesson has been boiled down to its most basic concepts, skills, and means of delivery, the teacher has the ability to adapt the lesson for any reason. If technology fails, if a school day is shortened because of a snow storm, if students are not grasping a concept, the teacher has the ability to seamlessly alter the plan to account for the changing situation.

The bottom line is that if we are expecting students to learn simpler, teachers must teach simpler. All instruction should be done in the least restrictive manner possible. This is especially true when it comes to technology in the classroom. Three questions must be asked:

1) Does this technology simplify instruction?
2) Is this technology easily accessible for all students involved?
3) Does this technology *support* rather than *replace* instruction?

If the answer to any of these questions is no, then this is likely not a technology that should be used.

Answers to the questions above will, of course, vary from school to school, and depend on the specific technology. There is, of course, a place in education for iPads, YouTube, Prezis, Google Docs, and myriad other technologies. However, these tools in and of themselves do not make for better teaching. In fact, as we have seen, more use of technology—inside and outside the classroom—can make it more difficult for students to learn and teachers to teach. When using these types of technologies, the focus of the lesson often is method of instruction (how to turn on the iPad, or the twisting and turning of the Prezi) rather than on the content or skills being taught.

Technology, by definition, is supposed to make life easier. Too frequently, educational technology can have the opposite effect. Educators look to add overcomplicated technological tools in order to meet the demand for increased tech in classrooms. It's simply technology for technology's sake. Recently, a friend of mine was telling me about how she had logged in to her daughter's online grade report and was shocked to see her daughter had received an F on an assignment for not returning a form that was supposed to be signed by a parent. She knew she had signed her daughter's form, so when she asked her about it, she was surprised by her daughter's response. "It must not have gone through," her daughter explained.

"'Gone through,' you mean you didn't hand it in?" my friend inquired. The daughter explained that rather than collect the forms from the students, the teacher had them take a photograph of their signed forms on their phones, then submit the pictures through a digital drop-box. Apparently, unbeknownst to the daughter, the file size of her photograph was too large to be submitted. She ended up having to reformat the picture to reduce its size on her computer (something her cell phone couldn't do) in order to submit it successfully. All of this raises the question, Why not just have the students hand in the forms in the first place? When technology creates more work than it saves, whether it's in a classroom or anywhere else, it should be discarded.

Our first S, delivering instruction in the simplest possible manner, is rooted in the way generations of students have learned and have been taught. That may sound like I am advocating for teachers standing and lecturing from bell to bell. After all, that is likely the simplest of all teaching methods and requires no tools of any kind. However, lecturing from bell to bell will not facilitate our second two S's: being *skills based* and *social*.

## Principle 2: Focus Instruction on Skills (What Students Are Able to *Do*)

Next let's take a look at why and how schools can focus on what students are able to do, versus on what they know. Of course students need content knowledge to make the best use of their skills. However, there is no significant value in memorizing the articles of the US Constitution, for example, without knowing how to apply those principles to life today. Whenever possible, lessons must focus on the skills students need to survive in our world. Many will say that education today, therefore, must be centered around technology, and students should be encouraged to interact with as much technology as possible. The reasoning behind this argument is that since so much of modern life is dependent on technology, schools must teach these technologies.

This might be an appealing idea, but it is folly. As I discussed earlier, students come to school already technologically dependent. They have been exposed to technological gadgets from birth, and have a difficult time navigating their world without them. Students need no help from schools in developing their tablet, smartphone, or Twitter skills. They are doing this on their own. What they need help with is critical thinking, problem-solving, and community building. They need to know how to analyze documents and literature. They need to know how to act. They need to know how to exercise. They need to know how to solve for unknowns. They need to know how to conduct experiments and speak other languages. They need to know how to analyze different investment vehicles. They need to know how to fix and build things. In short, students need to know *how to do* many, many things. This is the job of educators. In an assessment class I took a long time ago, the professor used the classic example of teaching a child to ride a bike. If you want to teach a child how to ride a bike, you don't give him a paper and pencil test on the parts of the bicycle. You don't have him watch a video on riding a bike. You teach the skill of riding the bike, and you assess whether or not the child has mastered the skill.

The spirit of this simple example should permeate our classrooms, from preschool on up, regardless of the course. Some disciplines—such as math, music, and English—lend themselves more easily to skills instruction. Others, such as history and civics, are more content heavy. That said, *all* courses can and should begin with the end in mind—*What do my students need to know how to* do? Knowledge for the sake of the knowledge is fine, but it is not likely to help our students reach their potential if they cannot *do* anything with the knowledge. That said, it is not enough to teach skills for skills sake, either. Just as the teacher who lectures from bell to bell and rationalizes the lesson by saying, "I'm teaching note-taking skills" does not fit into this model, neither does teaching Google Docs. Schools must teach skills that are relevant to the subject matter and important for our students' future. Transcribing the words coming out of a person's mouth at conversational speed is not a useful skill in a real-world sense unless you grow up to be a courtroom stenographer or a 1960s secretary. This is equally as useless as teaching students how to upload documents to a program that will probably not be around in two years, much less by the time they enter the workforce. The biggest fallacy in technology education is that by teaching students how to use a specific technology, you make them more tech savvy. Although that may be true in a limited sense specific to that one technology (that is likely to change in the blink of an eye), if we give students the more general ability to critically think and problem-solve, they will be much more capable of dealing with a variety of technologies and adapting to their particular circumstances.

What is unnerving about many advertisements for educational-technology tools is that they try to pass off skill *with the tool* as being the important skill. If you are teaching a course in how to use an iPad, then skill with an iPad is obviously critical. However, most classes today are not teaching courses in how to use specific gadgets. If you are teaching a Spanish class, the skills you teach must be Spanish skills. If students use a tool during their skills-acquisition journey, and it is a natural, simple part of the instruction, that is a bonus. However, teaching the gadget cannot replace teaching the content-related skill. Education should focus on skills that will never become unnecessary or outdated. These are the skills like reading, writing, arithmetic, critical thinking, and problem-solving. These are the skills that have been the focus of most education systems since the advent of educational systems.

Too often, technology-based education actually teaches technology use to the *detriment* of the skill students should be learning. For instance, many school systems now offer online PE courses. Yes, you read that correctly, virtual physical education, an oxymoron. Students are taught how to use digital heart-rate monitors to record and document their physical activity. They are expected to spend an hour a night in a target range of "beats per minute," indicating they were engaged in some type of strenuous activity. At the end of the week, they are expected to upload their data into an online exercise log. The problem is, students quickly learn how to upload the same heart-rate file over and over again. A student only has to exercise once in an entire semester and simply rename the file several dozen times. One student admitted he never even exercised once he just hooked his heart-rate monitor up to his dog for an afternoon. He got an A+ for the semester. The students of this class are no longer learning anything about healthy lifestyles, which is the intended purpose of physical education. They're instead learning how to copy digital heart-rate monitor files and paste them into a website they'll never see again.

One school in a suburban metropolitan area that I visited gave me the following example from a US government course. They had set a team goal that had to do with students knowing the basics of each article, section, and amendment of the US Constitution. That had been their goal for years, and it was taken for granted that having students know what the Constitution says is desirable and necessary. One year, though, they hired a new teacher who was fresh out of college. At a team meeting, he asked—purely for his own understanding—*why* it was important for students to know what the Constitution says. His question was met with silence followed by stammering and finally a discussion among team members. Oddly, no one could clearly articulate why it was important for students to understand details of the most essential and transformative document of the American experience. Finally, someone stumbled onto the beginnings of an answer: "Well, what if one of our students doesn't know her rights and gets stopped by a police officer? She can get in a lot of trouble if the officer oversteps legal bounds *and the kid doesn't know what to do.*" Bingo! What they went on to realize is that *knowing* rights, while important, is not enough. The student must be able to *apply* those rights to specific situations. They went on to augment their instruction. They still teach what the Constitution says—there is no way to know what it means if you do not know what it says. However, after they get through that instruction, they give teams of students different scenarios—some from Supreme Court cases, some from local news stories, and some from teachers' own experiences—and ask how the Constitution applies in each of the scenarios. The discussions at the conclusion of the unit are now much richer, and focus on what students, teachers, politicians, police officers, judges, and business owners should *do* given what the Constitution says.

If this sort of shift in focus—toward what students can do—can take place in a high school social studies course, it can take place in any course. In fact, this shift incorporates all three *S*'s: it is simple, it teaches students how to apply the Constitution to their lives, and it fosters social interaction.

### Principle 3: Foster Face-to-Face Social Interaction

This brings us to the final *S*. Schools need to help students redevelop seemingly lost *social skills*. We can do this by having students work in both small groups of their peers at times and larger groups when appropriate. By making them put down the screens and work face-to-face, students are developing critical and important skills. They are learning how to effectively communicate and read body language. It reinforces important social norms like what are appropriate and productive behaviors when trying to get something accomplished, and what types of comments will receive a positive response versus a negative one. It will instill in them a sense of varying appropriate relationships for their peers and their tribal chiefs (teachers and administrators).

All of these things are best accomplished through actual face-to-face interactions with others. Simply put, if students are on screens, they are not interacting with live human beings. If they are not interacting with live human beings, they are not building their school communities. One of the problems is that we are routinely told that there is a "digital community" to which students belong. However, "digital natives" are forgetting how to engage in actual face-to-face conversation. Encouragingly, though, when students are unplugged and detached from their screens, they are hungry for human interaction. Students devour well-planned, focused, high-value lessons that ask them to interact with one another. What is even more encouraging is that when students interact with one another in school, they continue to do so outside school. Having these relationships and being accountable to one another increase students' desire to learn, and, as a result, students achieve more. Relationships can be built and communities can thrive only when students are asked to work on relationship- and community-building skills. Fortunately, these skills are innate, and students are hungry to develop them.

Not long ago, teachers feared any "free time" in class because students would be loud and overenergetic. Today, however, free time has come to mean "screen time." What many teachers today observe in our classrooms is what has been reported to me by teacher after teacher—students are willing to engage in conversation and debate during a lesson. However, if a lesson ends four or five minutes early and teachers tell students they have a few minutes of free time, the default for most students is to isolate. The phones come out, the conversation stops, and the room falls silent. The need to reverse this trend goes beyond simply practicing conversation. As discussed earlier, schools today are battling teen depression and anxiety like never before. However, teachers are also giving students more screen time than ever before—which is likely one of the causes of the very thing they are trying to eradicate. Lessons that encourage social interaction and connection to community have never been more critical. Teachers must lead students back to their natural, social abilities. When students interact with one another in skills-focused lessons delivered in a simple manner, they thrive. When we teach simpler, learning becomes simpler and student achievement improves. Using these three *S*'s will help students and teachers recapture and reinvigorate learning.

## Introducing Kids to Their Own Brains

The three *S*'s are a great start, but they are not enough. For schools to be great today, we have to change the way we think about teaching, learning, and the role technology plays in both. Think of teaching as doing brain surgery. We need a completely sterile environment before we begin. The first step toward creating good thinkers is to create a pro-thinking environment in which they can thrive. Because the natural inclination of so many of our students today is to resort to multiswitching, teachers need to take extra precautions when trying to get students to focus. We try to think of the classroom as an operating room and treat distractions as pathogens. You can't begin surgery until you've cleared the room of all pathogens. Step one, of course, is making sure all phones are off and out of the way.

Once we've cleared the room of distractions, we can get to work. Yet many digiLearners today are attempting to work with a perfect stranger, their brain. It's almost impossible to work with strangers. In order to develop better critical thinkers, students must first develop a better relationship with their thoughts. To be truly effective, a pair or group working together must start by developing a working relationship. They must get to know one another, how they work and how they think. This is the reason so many classes begin with icebreakers and so many workplaces use team building activities. It takes time to build the intimate working dynamic that is necessary for any pair or group.

However, if you watch modern digiLearners, you can't help but come to the conclusion that they hate the sound of their own inner monologues. They fill any amount of mental downtime, no matter how brief, with the noise of other people's voices to drown out their own thoughts. Teachers see this many times every day. For example, a student has to go to the restroom. He leaves and as he's heading out the door he immediately reaches for the headphones stashed inconspicuously in his pocket for the three-minute excursion. Thinking for those three minutes, or hearing what is going on in his head, is unthinkable. Another example: a girl finishes a test. She turns in her paper, returns to her seat, and turns on a movie on her phone. Another: a student is given some work to do at his desk. The paper hits the desk and the earbuds go in his ears, because classwork time is music time. These are all examples of things that happen in classrooms every day. Thinking can be exhausting, and it can make you uncomfortable, so many students choose to avoid it at all costs. Modern devices make this avoidance really easy.

Although people can ignore their inner monologue for automatic brain functions like breathing, swallowing, or reciting their dates of birth, they cannot ignore it for higher-level skills such as critical thinking. In order to become better problem-solvers, students need to get used to the sound of their own thoughts. The sound of thoughts, and nothing else, needs to fill their heads.

Watching a student put in his earbuds as he walks twenty feet from his classroom to the water fountain and back, only to remove them thirty seconds later, became the genesis for a learning strategy I use daily. This very simple yet effective strategy is a mandated period of silence set aside during every class. It can come as a warm-up or a culminating activity. It can come in the middle of a lecture. I don't want to miss an opportunity to have the students think about something. If a student asks a great question, we might have an impromptu thinking session. The reflection might only go on for a minute. But whatever the problem is or how it comes up, I want students to have a little time to think about a problem or a critical thinking question. More important, I want to give them what they need most: silence.

When it comes to this strategy, teachers must be firm about getting rid of screens and earbuds in class or it won't work. Students use noise to drown out their thoughts. This conversation is between them and their brain and no one else. I'll even have students introduce themselves to their inner consciousness the first time we do this activity. They have to say aloud, "Hi, I'm ——, it's wonderful to meet you, brain," to which their inner monologues normally introduce themselves back.

Especially early on, students should write down their thoughts. This serves several purposes. First, this makes every student accountable for actually completing the assignment. Second, it gives students a view into the critical window connecting their subconscious brains and a conscious activity like writing. When students are tasked with writing down what they are thinking about, they get a very real example of how thinking works. Not only do they have thoughts in their brains, they have to think about those thoughts and then think about how to put them into words on a piece of paper. It's harder than it sounds, and it is a critical skill for students to develop.

## Encouraging Kids to Work at Learning

Once we've established a classroom that is conducive to critical thinking, we need to move on to step two, working at learning. Often, gyms post cliché motivational phrases on their walls to "inspire" their clientele. Walk into any weight room and you'll likely hear one meathead yell at another some variation of "You can do this!" or even worse, "Feel the burn!"

As coaches and teachers, we're not huge on reciting cliché mantras as a method of motivation, but on our school's weight room wall, someone painted, "No pain, no gain." As much as we might scoff at this banal phrase, there is quite a bit of truth in it. It brings back fond memories of an infomercial from the '90s for an electric ab machine. The user simply put the electrodes on his or her abs, turned on the machine, and voilà!—electric pulses would contract the user's muscles in a method similar to a crunch. You could be doing crunches anywhere, even from the comfort of your own couch while watching television and eating potato chips. It seemed like this technology would revolutionize working out. However, this fad quickly faded when people realized this technology (sad trombone: *womp, womp*) . . . well, it didn't work.

People have long accepted the truth that when it comes to improving one's body gain comes from pain. It's a hard truth. We don't like it, so these gimmicky gadgets will likely always be with us. However, the more one struggles to keep one's pace up during a run, or the more one struggles to get that last rep in a set of bench presses, the better the body will be for it.

If we are willing to accept this truth for our physique, why would we expect anything different for our social learning and brain development? When it comes to education, technology has been designed to make acquiring information easier. Students have access to everything from calculators to Google to instantly answer any question that may come to mind. Rather than struggle through the mental rigors of trying to solve $7 \times 8$, they can simply pull out their phones and let it answer for them.

Rather than using technology to improve their understanding, many digiLearners have grown to depend on it to solve their problems for them. Technology to them is used as a crutch rather than an aid. They can't answer why because Google has difficulty easily answering "why." They think in terms of what and who, and of facts that will easily pop up with the first Internet search. They can easily define because Google can easily define. Type in a question and *poof*, the answer appears. Technology has turned their brains into doughy lumps of goo rather than fit responsive brains. Perhaps more than anything, students must be able to solve problems, and we must teach them how to do this.

Therefore, what we want to do in schools is adopt the gym mentality of "No pain, no gain!" As simple as it may sound, in order to create better problem-solvers, we must first force students to solve their own problems. We have to create classroom climates that foster problem-solving skills and deep thinking. That cannot be done without hard work. We want the brain to do the heavy lifting here.

## Insist on Tech Distraction–Free Classrooms

Teachers need to insist that classrooms be technology-free zones from day one. Students know that staring at the glow emanating from their laps is strictly forbidden, as are laptops, iPads, e-readers, or anything that can be turned into a problem-solving crutch. Once this climate has been established, we can move on and forget the day-to-day combat. As parents, the more you insist on this to be a schoolwide rule, the more of a reality that can become.

However, if your school system is not going to accept a total prohibition of technology from classrooms, teachers can still designate any critical-thinking activity as technology-free time. Students find this invigorating. When teachers force students to take their brains off autopilot and fly solo, we also give them a much-needed break from Twitter, YouTube, and Snapchat.

One of the primary obstacles to this is that more and more schools are opting for online textbooks. We saw the effect this had on Brett in chapter 1, and again on a student in chapter 4. For some reason, though, it has been deemed acceptable or even "best practice" to give a student a textbook on a laptop computer for which she has to have a log-in code and password, have her wait while it boots up, direct her to a book publisher's website where she enters another log-in code and password, and instruct her to read a few pages from her online textbook. Alternatively, she could have just taken her actual textbook out of her actual backpack and done the same thing. Online textbooks tend to be cheaper for school systems, and they provide massive profit potential for producers of laptops and textbooks, but they often stand in the way of learning. Why do we make students jump through hoops to do their work?

In the last chapter, we discussed how online textbooks often don't live up to the promised benefits. However, there are more important reasons to keep these out of our schools. A 2013 *Science Nordic* article explains, "A new Norwegian study of 10th graders confirms that reading texts in print versus on a computer screen is better for some aspects of comprehension." Reading researcher Anne Mangen explains, "Numerous studies show that when you read a text on paper your understanding is deeper and longer lasting than if you read that same text on a computer."

The obvious solution is to give students actual textbooks. While this decision can be beyond the reach of the teacher or parent, school boards must understand that actual, physical textbooks are far simpler, yield better results, and are much more accessible to a much larger population. This is yet another issue where we must make our voices heard. When it comes to many supposed advances in education, we have outthought ourselves—far over-complicating things. At this point, we are not only reinventing "the wheel," we are reinventing "round."

To get to these textbooks, many schools that have not yet gone the route of giving each student his or her own device have "mobile computer labs" that are laptop carts teachers can check out and bring to their rooms for student use. In schools where these are used, the carts are a common resource, and they are often in disrepair. Either they do not get plugged in so the batteries are dead, or the school's network is down, or there are missing or stuck keys. They are also old. Distributing the laptops, getting them turned on, and getting all students to the right software application or website is guaranteed to take twenty or more minutes of class time. My experience and the interviews I have conducted indicate that, more often than not, these carts inhibit and slow teaching and learning. A poorly functioning mobile laptop cart does not simplify instruction, and it often results in restricted student access. Therefore, these carts fail the first and third S's because they are adding an unnecessary logistical hassle to instruction, and because once the students have the laptops they are isolated from one another. Students will tend to ignore the live human beings right next to them as they stare at their screens.

## Fostering Real Collaboration

Ed-tech advocates claim technology makes student collaboration easier, but we have rarely found that to be true. For example, if a lesson calls for students to collaborate at home using Google Docs, students without computers and Internet access are going to be frozen out. In an even more basic sense, however, Google Doc collaboration is usually unnecessary in the first place. Saying that students are "collaborating" with Google Docs is positively Orwellian. The use of Google Docs implies that students are working on a document from *separate locations*. We understand that this can be enormously helpful for people who live great distances from one another. However, our students are in *the same school*.

There are easy ways to simplify things. Students can meet *after school* in the same place, or they can all meet at someone's house.

This has the major advantage of having our students hone their social skills, which, if you have been in a classroom recently, you know could not be more critical. True "collaborating" is done face-to-face because it simplifies teaching and learning while further assisting in community building and social skill development.

## Using Technology to Support—Not Replace—Instruction

Some schools and school systems are going to dig in their heels. They are going to insist that teachers and students use digital technologies during the school day. Even so, there is a right way and a wrong way to use technology in the classroom. This became crystallized for me a few years ago. I was teaching AP economics. I had some of the brightest students in the school. Because they were so bright, I (foolishly) assumed a high level of knowledge, skill, and intuitive ability. To introduce the concept of "real" versus "nominal" gross domestic product, I found a YouTube clip that I thought was perfect (incidentally, "real" GDP is the inflation-adjusted dollar value of production of an economy, while "nominal" is the dollar value of production in a given year, without adjusting for inflation). It was from the movie *Spinal Tap*. It was a two-minute clip in which the band's lead guitarist, Nigel, was giving a reporter a tour of his guitars and equipment. In the clip, Nigel is very excited about the amplifier he has, because "it goes to eleven." He goes on to explain that most amps only go to ten, but this one is "one louder." The reporter asks why, if the amp is louder, he doesn't just make the top number ten. Nigel only stares blankly and says, "This one goes to eleven."

Students find this clip mildly amusing. I find it hilarious. At any rate, at the end of the clip I asked students which was the nominal measurement. Most students responded correctly, "eleven." In name ("nominally") the amp was, in fact, "one louder." However, of course, in a real sense the amp was no louder than if the top number were ten. That is, it was not *actually* any louder. Students said they understood it, so we moved on.

After class, however, a very bright, polite young lady approached me and said she had no idea what the point of the video was or how that was connected to gross domestic product. I gave her a much more complete explanation. She said she understood but that many of her classmates were lost. I asked them the next class if they understood, and most said that they had just guessed that eleven was nominal, or had a feeling for some reason that it was eleven, but they didn't understand why.

I then actually taught the concept and the connection between the two in more detail, and we all breathed a sigh of relief. I still use the video clip today, but I use it to introduce the concept and then I explain it and the connection. What I did initially was allow that technology tool to *replace* instruction. The way I use it now *supports* instruction, and there is obviously an enormous difference. By allowing the clip to replace my own instruction, I had complicated learning for my students. By changing the way I used it, I simplified my teaching. Learning a complex concept became simpler for my students. In my mind, this is the direction schools need to go: simpler, skills based, social.

For some reason, though, we continue to go in the opposite direction. For instance, suppose a high school English teacher has in his classroom a laptop and a projector and he has spent all night making a snazzy Prezi about the world of Jay Gatsby. This might look cool to another educator. However, at this point students have likely seen something like this about a thousand times before. Let's say, though, for the sake of argument that this one is the best of all time. That and five dollars will get the teacher a cup of coffee at Starbucks. But if that is all the lesson is—showing the presentation and lecturing from it—then what was the point of the technology? Ultimately, it is not any different than lecturing from notes at a podium. What if the Prezi had a twenty-minute video clip from YouTube of someone else lecturing about Gatsby? This person, however, uses funny sound effects and is wearing a purple jacket. Again, your technologically enhanced lesson is still, at best, not much better than lecturing from a podium. Further, suppose the presentation is stored on the school network, and when the teacher comes into school in the morning he learns that the network is down, or the projector is fried. What is the backup plan? If it is to "punt" and put in a DVD, then the lesson might have been too complicated.

A lesson that cannot be replicated or replaced if a piece of technology is unavailable is likely unnecessarily complex. That must be the starting point: What tools will lessons require, and will they be as effective if they are not available? If lessons will not be as effective without the tools, they are probably too complex. If they will be just as effective without the tools, then the technology wasn't needed in the first place. Very often, all that is needed to improve a lesson is to simplify it by removing unnecessary barriers between teacher and student.

Or consider this example from an article on HowStuffWorks.com about ten iPad apps for science education. The first app described is about the periodic table. "Open the app and instead of boring blocks with letters and numbers in them, you get an array of spinning objects that represent each element. Interested in bismuth? Touch the rotating crystal. Copper? Put your finger on the chain-link bracelet. The next screen adds extensive information about the element. You can drag your finger back and forth to change the direction and speed of the spinning item."

It goes on to say that if you buy some "inexpensive glasses" you can see even cooler pictures (and give more money to the developer). Doesn't it *sound* like students are *doing* something amazing? Many digiLearners would *love* this app. Essentially, though, by using this app, students are simply reading about the elements. There are clearly improved graphics over what a paper periodic table can offer. However, is that app high value enough to require the school system to purchase it and an iPad for every student? That seems unlikely. Students using that app are learning the skill of using the app, but they are not learning any new chemistry-related skills. They are just learning chemistry-related content. That can be simplified. Why introduce another piece of machinery to teach something that students already have in their textbooks and in the classroom?

In this same article, here is the description for an astronomy-related app: "Simply boot up, aim your iPad at the sky and, using its internal compass, it will tell you what constellations and stars you're looking at." This app has all types of other features that allow the user to see different planets and stars regardless of the user's location. Notice the difference here. With the earlier app about the periodic table, students were simply looking at a fancier periodic table. The app did not introduce any new skills, activities, or information compared to what a textbook could offer. However, this second app—pointing an iPad at the sky and getting a star map, day or night, given your specific position on the globe—does something impossible to do with paper star maps. Now students learn how to identify heavenly bodies in a much more direct, simple way than can be taught with a textbook. We would argue that this is an appropriate use of an iPad and an app. We don't think it's an ideal lesson, since it requires all students to have this particular gadget. However, it is a simple, skills-based use of a tool that cannot easily otherwise be done.

## Speaking Up

I'm not talking about creating a new network of schools with a radical change in philosophy. You don't need to get your fundraising friends together so that after 538 bake sales you'll have enough money to buy a small piece of land that will one day be where your great-grandchildren are finally educated. I'm talking about changing the schools we have into schools that are even better. I'd go so far as to call them ideal. The schools your kids attend have everything they need right now. This is about rallying community and faculty support to demand the schools our kids need.

As I have suggested before, PTAs are a great place to start. You're the *P*s. We're the *T*s. Together, we need to get our *A*s in gear if we're going to make necessary changes. Here are three fundamental questions parents should ask of their kids' teachers and school leaders:

- **Why is the digital version (of whatever the school is doing) better than the "analog" version?** You'll likely hear something along the lines of, "We have to prepare our kids for the future." At this point, you should know how to respond to that. Kids are already tech dependent. They know how to use their screens. What they need is to be able to think, focus, and build community. Ask for clarification about how digital tools help those things happen. If the answer you get is unsatisfactory, please share some of what you learned in this book. If the answer is satisfactory, please share it with me because I have never heard a satisfactory answer to this question.
- **Do you have any research showing the advantages of using screens for instruction?** The only studies I know of that show advantages show only that there is an increase in "student engagement" when students use screens. First of all, engagement (like many things in education) is tough

to quantify. More important, though, engagement in and of itself isn't necessarily a great thing. If we let our kids play video games all day long many of them will be completely absorbed and engaged in the game. That doesn't mean they're learning anything. However, ed-tech firms love to trumpet the "engagement" argument when selling their wares to schools. It's a red herring. There are at least as many studies showing the opposite about engagement.

- **May I opt my child out of screen-based instructional activities?** As I've noted, it may be impossible to avoid all screen time in school, at least in the near term. However, this is an important question to ask as it will give you a window into the school's philosophy about screens. If they seem shocked that anyone would want his or her child off screens, then you know you've got a battle on your hands. If, however, you are met with a thoughtful reply and an offer to collaborate on alternatives, you might already be close to that ideal school. In any event, you should know the landscape before you attempt to mobilize other parents in the community.

In summary, we must seek a world in which kids are taught how to think, focus, and successfully navigate social interactions. I cannot imagine anyone who cares about kids being opposed to those goals. The disagreement is about how to get there. Schools that focus on delivering instruction in a simple, skills-based, social manner are what we're after. We must demand schools that help students learn to think critically. Most of all, we must demand schools that understand that teaching and learning are hard work, and are not afraid of that work.

When we started teaching, like so many of our colleagues, we wanted to "change the world." However, we can only truly do this on the scale necessary if we have a tremendous amount of help. We hope you'll join us.

# Acknowledgments

F IRST AND FOREMOST, we would like to thank our wives and children. Without their support, patience, and love, this would never have been possible.

We would next like to thank our publisher, Chicago Review Press, and the wonderful Lisa Reardon, in particular. Their willingness to take a chance on the two of us and this vitally important idea has been both affirming and inspiring.

This would not have been possible without the help and support of Dr. Susan Greenfield, Dr. Richard Cytowic, and Dr. Richard Freed. Your willingness to help us spread this message is a testament to your commitment to the well-being of future generations.

Matt would like to thank his brother Chris Miles and his Uncle Ed Helder for their feedback and support.

Finally, we would like to thank the dozens of colleagues who have listened to us yammer on like street-corner doomsday preachers, and who were so willing to share their insights and expertise.

# Sources

## Chapter 2: The Myth of the Technology-Enhanced Superkid

Boerma, Lindsey. "Kids with Cell Phones: How Young Is Too Young?" *CBS News*, 2 Sept. 2014. Web. 22 Sept. 2015.

Cavalli, Earnest. "FCC Commish Blames *WoW* for College Dropouts." *Wired.com*, 11 Dec. 2008. Web. 1 Sept. 2016.

Freed, Richard. *Wired Child: Reclaiming Childhood in a Digital Age.* n.p., 12 Mar. 2015. Print.

Gentile, Douglas. "Pathological Video-Game Use Among Youth Ages 8 to 18: A National Study." *Psychological Science* 20.5 (2009): 594–602. Web.

Greenfield, Susan. *Mind Change: How Digital Technologies Are Leaving Their Mark on Our Brains.* New York: Random House, 2015. Print.

Hopson, John. "Behavioral Game Design." *Gamasutra*, 27 Apr. 2001. Web. 1 Sept. 2016.

Ito, Mizuko, Heather A. Horst, Matteo Bittanti, danah boyd, Becky Herr-Stephenson, Patricia G. Lange, C. J. Pascoe, Laura Robinson, Sonja Baumer, Rachel Cody, Dilan Mahendran, Katynka Z. Martinez, Dan Perkel, Christo Sims, and Lisa Tripp. "Living and Learning with New Media: Summary of Findings from the Digital Youth Project." *The John D. and Catherine T. MacArthur Foundation Reports on Digital Media and Learning.* Chicago: Macfound.org, Nov. 2008. Web. 8 Aug. 2016.

Kleinman, Alexis. "The Internet of Things: By 2020, You'll Own 50 Internet-Connected Devices." *Huffington Post*, 22 Apr. 2013. Web. 22 Sept. 2015.

Kosoff, Maya. "A California Couple Is in Prison for Neglecting Children While Playing *World of Warcraft.*" *Business Insider*, 11 Aug. 2014. Web. 1 Sept. 2016.

Kuchikomi. "The Figures Don't Lie: Smartphones Hurt Kids' Grades." *Japan Today RSS*, 6 Nov. 2015. Web. 31 Aug. 2016.

"Mobile Technology Fact Sheet." *Pew Research Center Internet, Science & Tech RSS*. Pew Research Center, 27 Dec. 2013. Web. 4 Dec. 2015.

Prensky, Marc. "Digital Natives, Digital Immigrants Part 1." *On the Horizon* 9.5 (2001): 1–6. Web. 15 July 2016.

———. "*Don't Bother Me Mom–I'm Learning!": How Computer and Video Games Are Preparing Your Kids for Twenty-First Century Success and How You Can Help!* St. Paul, MN: Paragon House, 2006. Print.

———. "*H. Sapiens Digital*: From Digital Immigrants and Digital Natives to Digital Wisdom." *Innovate: Journal of Online Education* February/March 5.3 (2009): n.p. Web.

Rideout, Vicky. "The Common Sense Census: Media Use by Tweens and Teens." *Commonsense.org*. Common Sense Media, 2015. Web. 25 Jan. 2016.

Rideout, Victoria J., Ulla G. Foehr, and Donald F. Roberts. *Generation M2: Media in the Lives of 8- to 18-Year-Olds*. Menlo Park, CA: Henry J. Kaiser Family Foundation, 2010. Web. 25 Jan. 2016.

Schoen, John W. "Goodbye, Empty Nest: Millennials Staying Longer with Parents." *CNBC*, 24 May 2016. Web. 29 Aug. 2016.

Selwyn, Neil. "The Digital Native—Myth and Reality." *AP Aslib Proceedings* 61.4 (2009): 364–379. Web.

Sheffield, Brandon. "GDC Europe: To Succeed in Free-to-Play, 'Exploit Human Weaknesses.'" *Gamasutra*, 18 Aug. 2010. Web. 17 July 2016.

"Study finds Average Age of Kids When They Get First Cellphone Is Six" *ABC7Chicago.com*, 7 Apr. 2015. Web. 22 Sept. 2015.

Swanson, Ana. "Why Amazing Video Games Could Be Causing a Big Problem for America." *Washington Post*, 23 Sept. 2016. Web. 2 Oct. 2016.

"Technology Addiction | Internet Addiction." *Addiction.com*, n.d. Web. 1 Sept. 2016.

Zimbardo, Philip, and Nikita Duncan. "The Demise of Guys." *Psychology Today*, 23 May 2012. Web. 30 Aug. 2016.

## Chapter 3: Reclaiming Your Child's Ability to Think

Adler, J. "Why Fire Makes Us Human." *Smithsonian Magazine*, 1 June 2013.

Aubusson, K. Internet Addiction Affects Brain. *Psychiatry Update*, 6 Mar. 2013.

Giang, Vivian. "The Science Behind How Boredom Benefits Creative Thought." *Fast Company*, 9 Feb. 2015. Web. 4 Aug. 2016.

"Harnessing Energy Sensory Processing." Retrieved 15 Dec. 2014 from www.zoneinworkshops.com/zonein-fact-sheet.html.

Haviland, W., D. Walrath, H. Prins, and B. McBride. *Evolution & Prehistory: The Human Challenge*. 9th ed., 193–195. Belmont, CA: Wadsworth/Cengage.

Keen, Andrew. *The Cult of the Amateur: How Today's Internet Is Killing Our Culture*. New York: Doubleday/Currency, 2007. Print.

Lin, F., Y. Zhou, Y. Du, L. Qin, Z. Zhao, and J. Xu. "Abnormal White Matter Integrity in Adolescents with Internet Addiction Disorder: A Tract-Based Spatial Statistics Study." *PLoS ONE* 7(1), 11 Jan. 2012 doi:10.1371/journal.pone.0030253.

McAuliffe, K. "If Modern Humans Are So Smart, Why Are Our Brains Shrinking?" *Discover Magazine*, 20 Jan. 2011.

"Mobile Technology Fact Sheet." *Pew Research Center's Internet & American Life Project RSS*, 2013. Web 15 Dec. 2014 from www.pewinternet.org/fact-sheets/mobile-technology-fact-sheet/.

OECD. "PISA 2012 Results: Creative Problem Solving: Students' Skills in Tackling Real-Life Problems." PISA 5, OECD Publishing 2014. doi: 10.1787/9789264208070-en.

Pearce, E., C. Stringer, and R. Dunbar. "New Insights into Differences in Brain Organization Between Neanderthals and Anatomically Modern Humans." *Proceedings of the Royal Society B: Biological Sciences*, 13 Mar. 2013. doi: 10.1098/rspb.2013.0168.

Rideout, Vicky. "The Common Sense Census: Media Use by Tweens and Teens." *Commonsense.org*. Common Sense Media, 2015. Web. 25 Jan. 2016.

Rideout, Victoria J., Ulla G. Foehr, and Donald F. Roberts. *Generation M2: Media in the Lives of 8- to 18-Year-Olds*. Menlo Park, CA: Henry J. Kaiser Family Foundation, 2010. Web. 25 Jan. 2016.

Selwyn, Neil. *Education and Technology: Key Issues and Debates*. London: Continuum International, 2011. Print.

Sherwood, C., F. Subiaul, and T. Zawidzki. "A Natural History of the Human Mind: Tracing Evolutionary Changes in Brain and Cognition." *Journal of Anatomy* 212 (2008): 426–454.

Weinberger, D. R., B. Elvevag, and J. N. Giedd. "The Adolescent Brain: A Work in Progress." National Campaign to Prevent Teen Pregnancy, June 2005. Web.

Yuan, Kai, Ping Cheng, Tao Dong, Yanzhi Bi, Lihong Xing, Dahua Yu, Limei Zhao, Minghao Dong, Karen M. von Deneen, Yijun Liu, Wei Qin, and Jie Tian. "Cortical Thickness Abnormalities in Late Adolescence with Online Gaming Addiction." *PloS ONE* 8(1), 9 Jan. 2013. doi: 10.1371/journal.pone.0053055.

## Chapter 4: Learning to Focus in the High-Tech World of Distraction

*The Big Lebowski*. Dir. Joel Coen and Ethan Coen. Polygram Filmed Entertainment, 1998. Film.

Flatow, Ira. "The Myth of Multitasking." *NPR*, 10 May 2013. Web. 25 Jan. 2016.

"Frequent Multitaskers Are Bad at It." *UNews*. University of Utah, 23 Jan. 2013. Web. 27 Jan. 2016.

Head, Alison. "May I Have Your Attention? The Brain, Multitasking, and Information Overload." Project Information Literacy, 12 Oct. 2011. Web. 17 Feb. 2016.

Hopson, John. "Behavioral Game Design." *Gamasutra*, 27 Apr. 2001. Web. 9 July 2016.

Kahneman, Daniel. *Thinking, Fast and Slow*. New York: Farrar, Straus and Giroux, 2011. Print.

Kosoff, Maya. "A California Couple Is in Prison for Neglecting Children While Playing *World of Warcraft*." *Business Insider*, 11 Aug. 2014. Web. 9 July 2016.

Loh, Kep Kee, and Ryota Kanai. "Higher Media Multi-tasking Activity Is Associated with Smaller Gray-Matter Density in the Anterior Cingulate Cortex." *PLoS ONE* 9 (2014). doi: 10.1371/journal.pone.0106698.

Lohr, Steve. "Slow Down, Brave Multitasker, and Don't Read This in Traffic." *New York Times*, 25 Mar. 2007. Web. 10 Feb. 2016.

Mark, Gloria, Daniela Gudith, and Ulrich Klocke. "The Cost of Interrupted Work: More Speed and Stress." *CHI '08 Proceedings of the SIGCHI Conference on Human Factors in Computing Systems* (2008): 107–110. Print.

National Highway Traffic Safety Administration US Department of Transportation. NHTSA'S Office of Behavioral Safety Research, Apr. 2012. Web. 26 Jan. 2016.

Rideout, Vicky. "The Common Sense Census: Media Use by Tweens and Teens." *Commonsense.org*. Common Sense Media, 2015. Web. 25 Jan. 2016.

Rideout, Victoria J., Ulla G. Foehr, and Donald F. Roberts. *Generation M2: Media in the Lives of 8- to 18-Year-Olds*. Menlo Park, CA: Henry J. Kaiser Family Foundation, 2010. Web. 25 Jan. 2016.

Rosen, Christine. "The Myth of Multitasking." *The New Atlantis* 20 (2008): 105–110. Web. 27 Jan. 2016.

Rubinstein, Joshua S., David E. Meyer, and Jeffrey E. Evans. "Executive Control of Cognitive Processes in Task Switching." *Journal of Experimental Psychology* 27.4 (2001): 763–797. Web. 26 Jan. 2016.

Sana, Faria, Tina Weston, and Nicholas J. Cepeda. "Laptop Multitasking Hinders Classroom Learning for Both Users and Nearby Peers." *Computers & Education* 62 (2013): 24–31. Web. 27 Jan. 2016.

Simons, Daniel J., and Christopher F. Chabris. "Gorillas in Our Midst: Sustained Inattentional Blindness for Dynamic Events." *Perception* 28.9 (1999): 1059–1074. Web. 25 Jan. 2016.

Strayer, David L., Jason M. Watson, and Frank A. Drews. "Cognitive Distraction While Multitasking in the Automobile." *The Psychology of Learning and Motivation: Advances in Research and Theory* (2011): 29–58. Web. 27 Jan. 2016.

University of California, Los Angeles. "Multi-tasking Adversely Affects Brain's Learning, UCLA Psychologists Report." *ScienceDaily*, 26 July 2006.

Watson, Jason M., and David L. Strayer. "Supertaskers: Profiles in Extraordinary Multitasking Ability." *Psychonomic Bulletin & Review* 17.4 (2010): 479–485. Web. 25 Jan. 2016.

## Chapter 5: Escaping the Digital World of Anxiety

Anderson, Nick. "Applied to Stanford or Harvard? You Probably Didn't Get In. Admit Rates Drop, Again." *Washington Post*, 1 Apr. 2016. Web. 24 Aug. 2016.

"Average High School GPAs Increased Since 1990." *U.S. News and World Report*, 19 Apr. 2011. Web. 22 Aug. 2016.

Bichell, Rae Ellen. "Suicide Rates Climb in U.S., Especially Among Adolescent Girls." *NPR*, 22 Apr. 2016. Web. 22 Aug. 2016.

Caplan, Scott E. "Relations Among Loneliness, Social Anxiety, and Problematic Internet Use." *CyberPsychology & Behavior* 10.2 (2007): 234–242. Web.

Collins, Sam P. K. "Americans Are More Depressed Than They've Been in Decades." *Medium*. ThinkProgress, 2 Oct. 2014. Web. 22 Aug. 2016.

Freed, Richard. *Wired Child: Reclaiming Childhood in a Digital Age*. n.p., 12 Mar. 2015. Print.

Gentile, Douglas. "Pathological Video-Game Use Among Youth Ages 8 to 18: A National Study." *Psychological Science* 20.5 (2009): 594–602. Web.

Hamer, M., E. Stamatakis, and G. Mishra. "Psychological Distress, Television Viewing, and Physical Activity in Children Aged 4 to 12 Years." *National Center for Biotechnology Information*. Pediatrics, May 2009. Web. 23 Aug. 2016.

Leménager, Tagrid, Julia Dieter, Holger Hill, Sabine Hoffmann, Iris Reinhard, Martin Beutel, Sabine Vollstädt-Klein, Falk Kiefer, and Karl Mann. "Exploring the Neural Basis of Avatar Identification in Pathological Internet Gamers and of Self-Reflection in Pathological Social Network Users." *Journal of Behavioral Addictions* (2016): 485–499. Web.

Loveless, Tom. "Homework in America." *Brookings.* The Brookings Institution, 18 Mar. 2014. Web. 23 Aug. 2016.

Luhrmann, T. M. "Is the World More Depressed?" *New York Times*, 24 Mar. 2014. Web. 22 Aug. 2016.

Martin, Karen, Juan Castro, Anil Saini, Manzoor Usman, and Dale Peeples. "Electronic Overload: The Impact of Excessive Screen Use on Child and Adolescent Health and Wellbeing." Perth: Department of Sport and Recreation. Aug. 2011. Web 23 Aug. 2016.

Messias, Erick, Juan Castro, Anil Saini, Manzoor Usman, and Dale Peeples. "Sadness, Suicide, and Their Association with Video Game and Internet Overuse Among Teens: Results from the Youth Risk Behavior Survey 2007 and 2009." *Suicide and Life-Threatening Behavior* 41.3 (2011): 307–315. Web. 23 Aug. 2016.

Odacı, Hatice, and Melek Kalkan. "Problematic Internet Use, Loneliness and Dating Anxiety Among Young Adult University Students." *Computers & Education* 55.3 (2010): 1091–1097. Web.

"Parenting in America: Children's Extracurricular Activities." *Pew Research Center's Social & Demographic Trends Project RSS*, 17 Dec. 2015. Web. 23 Aug. 2016.

Rosen, L. D., A. F. Lim, J. Felt, L.M. Carrier, N. A. Cheever, J. M. Lara-Ruiz, J. S. Mendoza, and J. Rokkum. "Media and Technology Use Predicts Ill-Being Among Children, Preteens and Teenagers Independent of the Negative Health Impacts of Exercise and Eating Habits." *Computers in Human Behavior.* U.S. National Library of Medicine, June 2014. Web. 23 Aug. 2016.

Shapiro, T. Rees. "Harvard-Stanford Admissions Hoax Becomes International Scandal." *Washington Post*, 19 June 2015. Web. 24 Aug. 2016.

Tang, J. C. M., and M. G. Livingston. "Correlation Between Facebook
    Usage and Loneliness and Depression." *Digitalcommons.csbsju.edu.*
    College of Saint Benedict and Saint John's University, 16 June 2012.
    Web. 23 Aug. 2016.

Tavernise, Sabrina. "U.S. Suicide Rate Surges to a 30-Year High." *New
    York Times*, 22 Apr. 2016. Web. 22 Aug. 2016.

Urist, Jacoba. "Is College Really Harder to Get into Than It Used to Be?"
    *Atlantic*, 4 Apr. 2014. Web. 24 Aug. 2016.

Van Den Eijnden, Regina J. J. M., Jeroen S. Lemmens, and Patti M.
    Valkenburg. "The Social Media Disorder Scale" *Computers in Human
    Behavior* 61 (2016) 478–487. Web. 23 Aug. 2016.

Wolfgang, Ben. "Number of High-School Students with Jobs Hits 20-Year
    Low." *Washington Times*, 24 May 2012. Web. 23 Aug. 2016.

Wong, Alia. "The Activity Gap." *Atlantic*, 30 Jan. 2015. Web. 23 Aug.
    2016.

## Chapter 6: Reestablishing Support from Home

Freed, Richard. "Back to Basics: Raising Children in the Digital Age."
    *Huffington Post*, 14 Aug. 2016. Web. 21 Sept. 2016.

———. *Wired Child: Reclaiming Childhood in a Digital Age.* n.p., 12 Mar.
    2015. Print.

Gentile, Douglas A., Rachel A. Reimer, Amy I. Nathanson, David A.
    Walsh, and Joey C. Eisenmann. "Protective Effects of Parental
    Monitoring of Children's Media Use." *JAMA Pediatrics* 168.5 (2014):
    479–484. Web.

Hoskins, Rob. "A Father's Day Resolution: Become an Emotional,
    Involved Dad." *Huffington Post*, 10 June 2011. Web. 21 Sept. 2016.

Howard, Jacqueline. "Americans Devote More Than 10 Hours a Day to
    Screen Time, and Growing." *CNN*, 29 July 2016. Web. 21 Sept. 2016.

Margalit, Liraz. "What Screen Time Can Really Do to Kids' Brains."
    *Psychology Today*, 17 Apr. 2016. Web. 21 Sept. 2016.

Mesch, Gustavo S., and Ilan Talmud. *Wired Youth: The Social World of
    Adolescence in the Information Age.* London: Routledge, 2010. Print.

Nolte, Dorothy, and Rachel Harris. *Children Learn What They Live: Parenting to Inspire Values.* New York: Workman, 1998. Print.

Oberschneider, Michael, and Guy Wolek. *Ollie Outside: Screen-Free Fun.* Golden Valley, MN: Free Spirit, 2016. Print.

Prooday, Victoria. "Why Are Our Children So Bored at School, Cannot Wait, Get Easily Frustrated and Have No Real Friends?" *YourOT.com.* n.p., 16 May 2016. Web. 21 Sept. 2016.

Richards, Rosalina, Rob McGee, Sheila M. Williams, David Welch, and Robert J. Hancox. "Adolescent Screen Time and Attachment to Parents and Peers." *Archives of Pediatrics & Adolescent Medicine* 164.3 (2010): 258–262. Web.

## Chapter 7: Revitalizing Social Interaction

Ahnert, Lieselotte, Anne Milatz, Gregor Kappler, Jennifer Schneiderwind, and Rico Fischer. "The Impact of Teacher–Child Relationships on Child Cognitive Performance as Explored by a Priming Paradigm." *Developmental Psychology* 49.3 (2013): 554–567. Web.

Fredrickson, Barbara L. "Your Phone vs. Your Heart." *New York Times,* 23 Mar. 2013. Web. 25 Sept. 2016.

Freed, Richard. "Parent Like a Tech Exec." *RichardFreed.com,* 16 Sept. 2014. Web. 25 Sept. 2016.

Gehlbach, Hunter, Maureen E. Brinkworth, Aaron M. King, Laura M. Hsu, Joseph McIntyre, and Todd Rogers. "Creating Birds of Similar Feathers: Leveraging Similarity to Improve Teacher-Student Relationships and Academic Achievement." *Journal of Educational Psychology* 108.3 (2016): 342–352. Web.

Gentile, D. A., H. Choo, A. Liau, T. Sim, D. Li, D. Fung, and A. Khoo. "Pathological Video Game Use Among Youths: A Two-Year Longitudinal Study." *Pediatrics* 127.2 (2011): n.p. Web.

Gilkerson, Luke. "Get the Latest Pornography Statistics." *Covenant Eyes,* 19 Feb. 2013. Web. 25 Sept. 2016.

Gordon, Serena. "Could Excess Computer, TV Time Harm Kids Psychologically?" *HealthDay,* 11 Oct. 2010. Web. 24 Sept. 2016.

Grasgreen, Allie. "Survey Suggests Technology Harms Students Socially." *Insidehighered.com*, 21 Mar. 2013. Web. 24 Sept. 2016.

Howard, Jacqueline. "Children May Be Losing Their Ability to Read Emotions, but There's a Fix." *Huffington Post*, 26 Aug. 2014. Web. 25 Sept. 2016.

Kühn, Simone, and Jürgen Gallinat. "Brain Structure and Functional Connectivity Associated with Pornography Consumption." *JAMA Psychiatry* 71.7 (2014): 827–834. Web.

Leménager, Tagrid, Julia Dieter, Holger Hill, Sabine Hoffmann, Iris Reinhard, Martin Beutel, Sabine Vollstädt-Klein, Falk Kiefer, and Karl Mann. "Exploring the Neural Basis of Avatar Identification in Pathological Internet Gamers and of Self-Reflection in Pathological Social Network Users." *Journal of Behavioral Addictions* (2016): 485–499. Web.

Ling, Richard Seyler. *The Mobile Connection: The Cell Phone's Impact on Society*. San Francisco: Morgan Kaufmann, 2004. Print.

Merchant, Safiya. "Lakeview High School to Ban Cell Phones in Class." *Battle Creek Enquirer*, 21 July 2016. Web. 25 Sept. 2016.

Page, Angie S., Ashley R. Cooper, Pippa Griew, and Russell Jago. "Children's Screen Viewing Is Related to Psychological Difficulties Irrespective of Physical Activity." *Pediatrics* 126.5 (2010): n.p. Web.

Rideout, Victoria J., Ulla G. Foehr, and Donald. F. Roberts. *Generation M2: Media in the Lives of 8- to 18-Year-Olds*. Menlo Park, CA: Henry J. Kaiser Family Foundation, 2010. Web. 25 Jan. 2016.

Turkle, Sherry. *Alone Together: Why We Expect More from Technology and Less from Each Other*. Philadelphia: Basic Books, 2011. Print.

Turkle, Sherry. *Reclaiming Conversation: The Power of Talk in a Digital Age*. New York: Penguin, 2015. Print.

## Chapter 8: Technology Is Widening—Not Closing—the Achievement Gap

"Bill Gates Keeps Close Eye on Kids' Computer Time." *Reuters*, 20 Feb. 2007. Web. 8 Aug. 2016.

Bilton, Nick. "Steve Jobs Was a Low-Tech Parent." *New York Times*,
    10 Sept. 2014. Web. 8 Aug. 2016.

Erickson, Kirk I., Charles H. Hillman, and Arthur F. Kramer. "Physical
    Activity, Brain, and Cognition." *Current Opinion in Behavioral
    Sciences* 4 (Aug. 2015): 27–32. Web. 8 Aug. 2016

Freed, Richard. *Wired Child: Reclaiming Childhood in a Digital Age.* n.p.,
    12 Mar. 2015. Print.

Gehlbach, Hunter, Maureen E. Brinkworth, Aaron M. King, Laura
    M. Hsu, Joseph McIntyre, and Todd Rogers. "Creating Birds of
    Similar Feathers: Leveraging Similarity to Improve Teacher–Student
    Relationships and Academic Achievement." *Journal of Educational
    Psychology* 108.3 (2016): 342–352. Web.

Gentile, Douglas. "Pathological Video-Game Use Among Youth Ages 8
    to 18: A National Study." *Psychological Science* 20.5 (2009): 594–602.
    Web.

Jenkin, Matthew. "Tablets Out, Imagination In: The Schools That Shun
    Technology." *Guardian*, 2 Dec. 2015. Web. 8 Aug. 2016.

Ko, Chih-Hung, Gin-Chung Liu, Sigmund Hsiao, Ju-Yu Yen, Ming-Jen
    Yang, Wei-Chen Lin, Cheng-Fang Yen, and Cheng-Sheng Chen.
    "Brain Activities Associated with Gaming Urge of Online Gaming
    Addiction." *Journal of Psychiatric Research* 43.7 (2009): 739–747. Web.

Murphy Paul, Annie. "Educational Technology Isn't Leveling the Playing
    Field." *Slate*, 25 June 2014. Web. 25 Sept. 2016.

Reardon, Sean F. "The Widening Income Achievement Gap." *Educational
    Leadership: Faces of Poverty.* ASCD, May 2013. Web. 8 Aug. 2016.

Richtel, Matt. "A Silicon Valley School That Doesn't Compute." *New York
    Times*, 22 Oct. 2011. Web. 29 Aug. 2016.

Rideout, Vicky. "The Common Sense Census: Media Use by Tweens and
    Teens." *Commonsense.org.* Common Sense Media, 2015. Web. 25 Jan.
    2016.

Rideout, Victoria J., Ulla G. Foehr, and Donald F. Roberts. *Generation
    M2: Media in the Lives of 8- to 18-Year-Olds.* Menlo Park, CA: Henry
    J. Kaiser Family Foundation, 2010. Web. 25 Jan. 2016.

Schwarz, Alan, and Sarah Cohen. "A.D.H.D. Seen in 11% of U.S. Children as Diagnoses Rise." *New York Times*, 31 Mar. 2013. Web. 30 Aug. 2016.

Swing, E. L., D. A. Gentile, C. A. Anderson, and D. A. Walsh. "Television and Video Game Exposure and the Development of Attention Problems." *Pediatrics Online* (2010): 126, 214. Web.

"Technology Can Close Achievement Gaps, Improve Learning." *Stanford Graduate School of Education*, 10 Sept. 2014. Web. 25 Sept. 2016.

Vigdor, Jacob L., Helen Ladd F., and Erika Martinez. "Scaling the Digital Divide: Home Computer Technology and Student Achievement." *Economic Inquiry* 52.3 (2014): 1103–1119. Web.

Whitmire, Richard, and Susan McGee Bailey. "Gender Gap." *EducationNext.org*, 15 Jan. 2010. Web. 30 Aug. 2016.

Zimbardo, Philip, and Nikita Duncan. "The Demise of Guys." *Psychology Today*, 23 May 2012. Web. 30 Aug. 2016.

Zimbardo, Philip, and Nikita Duncan. "Why Society Is Failing Young Boys." *Huffington Post*, 25 May 2012. Web. 30 Aug. 2016.

## Chapter 9: The Education-Industrial Complex

"Attend." *EdNET 2016*, n.d. Web. 19 Sept. 2016.

Barshay, Jill. "Why a New Jersey School District Decided Giving Laptops to Students Is a Terrible Idea." *Hechinger Report*. Teachers College at Columbia University, 29 July 2014. Web. 19 Sept. 2016.

Campbell, Mikey. "Los Angeles School District to Shift Away from Apple's iPad to Windows, Chromebook." *AppleInsider*, 30 June 2014. Web. 19 Sept. 2016.

Cherry, Bobby. "Quaker Valley Replacing 490 Broken, 1-Year-Old Laptops." *TribLIVE.com*. 27 Aug. 2014. Web. 19 Sept. 2016.

Gilbertson, Annie. "The LA School iPad Scandal: What You Need to Know." *NPR*, 27 Aug. 2014. Web. 19 Sept. 2016.

"Findings." *One-to-oneinstitute.org*, n.d. Web. 18 Sept. 2016.

Fredrickson, Scott, Phu Vu, and Sherry R. Crow. "Availability and Use of Digital Technologies in P-12 Classrooms of Selected Countries." *Issues and Trends in Educational Technology* 2.1 (2014): 1–14. Print.

Husock, Howard. "Bill Gates and the Common Core: Did He Really Do Anything Wrong?" *Forbes*, 18 June 2014. Web. 19 Sept. 2016.

Kardaras, Nicholas. "Screens in Schools Are a $60 Billion Hoax." *Time*, 31 Aug. 2016. Web. 19 Sept. 2016.

Layton, Lyndsey. "How Bill Gates Pulled Off the Swift Common Core Revolution." *Washington Post*, 7 June 2014. Web. 19 Sept. 2016.

Mencimer, Stephanie. "Rupert Murdoch Compares US Education System to Third World Country's." *Mother Jones*, 14 Oct. 2011. Web. 18 Sept. 2016.

Peterson, Andrea. "Google Is Tracking Students as It Sells More Products to Schools, Privacy Advocates Warn." *Washington Post*, 28 Dec. 2015. Web. 17 Sept. 2016.

Picciano, A., and J. Spring. *The Great American Education-Industrial Complex: Ideology, Technology, and Profit.* London: Routledge and Taylor & Francis Group, 2013. Print.

"POP1 Child Population: Number of Children (in Millions) Ages 0–17 in the United States by Age, 1950–2015 and Projected 2016–2050." *Childstats.gov.* Federal Interagency Forum on Child and Family Statistics, n.d. Web. 18 Sept. 2016.

Schaffhauser, Dian. "Report: Education Tech Spending on the Rise." *THE Journal*, 19 Jan. 2016. Web. 19 Sept. 2016.

Shi, Audrey. "Here Are the 10 Most Profitable Companies." *Fortune*, 8 June 2016. Web. 17 Sept. 2016.

Thorner, Nancy. "Gates and Pearson Partner to Reap Tens of Millions from Common Core." *Freedom Pub*, 25 Nov. 2014. Web. 19 Sept. 2016.

Villapaz, Luke. "How Google Took Over the American Classroom and Is Creating a Gmail Generation." *International Business Times*, 14 Aug. 2014. Web. 19 Sept. 2016.

Yu, Roger. "Technology, Costs, Lack of Appeal Slow E-Textbook Adoption." *USA Today*, 17 Jan. 2012. Web. 19 Sept. 2016.

## Chapter 10: Ideal Education in a Modern World

Christensen, Arnfinn. "Paper Beats Computer Screens." *ScienceNordic. com*, 13 Mar. 2013. Web. 21 Sept. 2016.

Franco, Michael. "10 iPad Apps for Teaching Kids About Science." *HowStuffWorks.com*, 8 Aug. 2011. Web. 21 Sept. 2016.

Mangen, Anne, Bente R. Walgermo, and Kolbjørn Brønnick. "Reading Linear Texts on Paper versus Computer Screen: Effects on Reading Comprehension." *International Journal of Educational Research* 58 (2013): 61–68. Web.

# Index